ZINFANDEL: A CROATIAN-AMERICAN WINE STORY
by JASENKA PILJAC

Associate editor
Ante Piljac & Višnja Piljac

Author and managing editor
Jasenka Piljac

Cover page drawing
Nikola Šimunić

Cover page design
Forma Ultima
Zlatko Rebernjak
Novi Goljak 19, Zagreb, Croatia

Photographs
Boris Kragić – Magenta studio, Split
Jasenka Piljac – personal archives

Reviewers
Professor Charles L. Sullivan
Professor of European history and American wine writer

Professor Carole P. Meredith
Professor Emerita, University of California at Davis

Professor Sibila Jelaska
Professor, Faculty of Science, University of Zagreb
Member of The Croatian Academy of Science and Art

Language editor
Professor Mary Anne Saunders
Associate Dean of Special and International Programs
The George Washington University

Graphical preparation
AGMAR d.o.o
Zagreb, Croatia

Print/Publisher
Zrinski d.d.
Ivana Novaka 13, Čakovec, Croatia

ISBN 953-155-081-6
Original in English

The publication of "Zinfandel: A Croatian-American Wine Story" has been financially supported by the Ministry of science, education and sport, the Ministry of culture of Croatia and the Croatian academy of arts and science.

No part of this publication may be reproduced, copied or used in any manner without the written consent of the author, Jasenka Piljac©. All rights reserved.

ZINFANDEL: A CROATIAN-AMERICAN WINE STORY

by
JASENKA PILJAC

with contributions from
ŽELIMIR BAŠIĆ and VLADIMIR JELASKA

*To the memory of my grandfather
whose favorite saying was ...
"In vino veritas"*

Loza*

Čvoraste su tvoje ruke
Prije mnogih ovdje bile
Kamena i strma polja
Čvrstim listom obavile

A i starac otac starcu
Tebe pamti, zvijezdam' kuje
Beštiju i bolest kune
Što ti žile i plod truje

Dok ti bura lomi grane
I nevera dok ne stane
Uz tri kapi dobrog vina
Gazda tvoj sad slavi sina

A na rivi malo niže
U konobi onkraj mora
Svijet se skuplja, čaše diže
I uz pjesmu stoljeć' staru
Nazdravlja u tvoju slavu!

Vine

Many months and years before
Your knotty hands have touched this floor
Climbing up the dry stone walls
Your seeds set life with each their fall

And all the settlers from miles around
Bless your grapes and kiss the ground
They curse the wind and nasty pests
That won't ever give you rest

While the storms rage more and more
In a small house at the seaside shore
A newborn's life is blessed with wine
For a long and happy child's lifeline

And at the tavern just below
The vintners finally take it slow
Tired from the work and land
They reach out one coarse hand
And raise the glasses to your health
Dreaming of their hard-earned wealth!

* Written and translated by J. Piljac

CONTENTS

FOREWORD	XI
ON THE DISCOVERY OF ZINFANDEL IN CROATIA	XIII
A WORD FROM THE AUTHOR	XV
ACKNOWLEDGEMENTS	XVI
WINEGROWING IN CROATIA - THE LEGACY OF OUR FOREFATHERS	1
Tracing back the origins of the vine	1
Grape-growing tradition in Croatia	4
Croatia's viticultural treasure	8
Wine in Croatian folk customs	11
Significant periods in Dalmatia's viticultural history	16
The antic period	16
Greek influence	17
Roman influence	20
The medieval period	22
The rises and falls of the 19th century	24
Dalmatian karst – a breeding ground for wine grapes	29
Geological composition of karst	31
Terra rossa	32
Climatic characteristics of karst	32
THE STORY OF ZINFANDEL	37
Biological characteristics of the Zinfandel vine	37
The Zinfandel wine	37
Zinfandel's popularity	39
Mysterious origins - the American view	41
A chronology of the European origins of Zinfandel	42
A nobleman's grape?	42
The trip to America	45
First modern traces in Italy	46
Zinfandel in Croatia	48
The Plavac mali controversy	53
Searching the vineyards of native Croatian varieties	67
Plavac mali – the child of Zinfandel	69
Relatedness of Zinfandel to other Croatian cultivars	70
Chances of finding Zinfandel in other countries	70
Kaštel Novi – the home of Crljenak and Zinfandel	72

The vineyard that will go down in history	73
Continuing the search	75
A wine road across the Atlantic	77

VITICULTURAL TRADITION OF THE TROGIR-KAŠTELA SURROUNDINGS 79

The Illyrian period	80
The Greeks in the Kaštela field	81
The development of viticulture under Roman rule	82
The rule of Croatian dukes	87
Viticulture and winemaking in the free city of Trogir	90
Trogir and Kaštela under the rule of Venetians and Turks	92
Winegrowing in Kaštela in the new era	96
Oidium, peronospera and phylloxera in the vineyards of the Trogir-Kaštela region	99
The wine clause and problems of the local winegrowers in the 19th century	105
Vineyard restoration and the establishment of nurseries in Kaštela	106
The assortment of grape varieties in the Kaštela field in the past	108
The absence of wine surpluses in the Kaštela field	112

A SPECIAL KIND OF A WINE STORY 117

NOTES 123
SELECT BIBLIOGRAPHY 131
FACTS ABOUT CROATIA AND WINE 134
BIOGRAPHY 136

FOREWORD

Jagoda Bush
Broadcast Journalist, Voice of America, Croatian Language Service

Having just completed a year-long series of interviews, for VOA Croatian, with key producers of Zinfandel throughout California, a series prompted by the 2001 discovery of Zinfandel's Croatian roots, I realized what a keen and enormous interest exists among all of them, growers and winemakers, for the history of Zinfandel and for Croatia, the land of its origin.

There have been a few other books about California's favorite grape, written by American authors, and covering mostly the grape's recent history, its 150 years long presence in the United States. What Dr. Piljac has now and very successfully done, with her "Zinfandel: A Croatian-American Wine Story," is giving us, for the first time in English language, a comprehensive, extremely well-researched history of Zinfandel in Croatia, in particular, and long and rich wine history of Croatia, in general. For American readers and wine lovers, among whom there is an astonishing number of Zinfandel devotees, a welcome and long-needed book, indeed!

What makes Dr. Piljac's book even more special and unique is the fact that she happened to be at the right place, at the right time, when the final chapter of the mystery of Zinfandel was being unraveled. Having a good and rare fortune of working closely with the best grape geneticist in the world, Dr. Piljac was thus able to give us the full scientific account, as well, of the Zinfandel story. Her vivid and fine descriptions of the events that led to the final discovery read as a good detective story, good as any, and her precious diary of trips made, with Prof. Carole Meredith, to the Dalmatian islands and other coastal locations, will serve, to the American readers, as the best promotion of Croatia, its unspoiled beauty, its rich history, and its abundance of unique wine.

Equally attractive the book ought to be to a Croatian reader, whether he/she is a history buff, or interested in popular science, or just a plain wine drinker, certainly a vine grower and a winemaker. With its marvelous insight into the popularity of Zinfandel in America, and the important place it occupies in the wine-industry of California, Dr. Piljac's book will inform and make the Croatian readers aware of the wonderful and potentially significant opportunities the discovery of Zinfandel's Croatian roots opens up for their land, reminding them, at the same time, of their own rich and proud viticulture heritage.

A true Croatian-American success story! Dr. Piljac's book is an invaluable and praiseworthy contribution to the existing literature on Zinfandel, the most popular grape in California, and to the existing literature on Croatia and its winemaking past, as well.

April 1, 2004.
Arlington, Virginia

ON THE DISCOVERY OF ZINFANDEL IN CROATIA:

One of the most important contributions was made by a Croatian student at UC Davis. Jasenka Piljac was a dishwasher in Meredith's laboratory in the early '90s. After graduating from Davis and returning to Zagreb, Piljac served as translator and research assistant during Meredith's 1998 sleuthing mission. "The timing worked out very well," says Meredith. "I would not have been able to do it without her. And that's an example of the almost eerie way things have fallen into place on this quest."

(A qoute from the July 10th, 2002, issue of the Los Angeles Times */ Interview with Professor Carole Meredith of UC Davis after the discovery of Crljenak kaštelanski – the Croatian match for Zinfandel)*

A WORD FROM THE AUTHOR

It has been widely accepted that *Vitis Vinifera* L. is not indigenous to the Americas and that all classic wine grapes have their roots somewhere in Europe and western Asia.

With the aid of modern day genetics, Professor Carole Meredith of the University of California at Davis has traced back the origins of numerous well-known and economically important cultivars of today, such as Cabernet Sauvignon, Chardonnay and Petite Sirah. However, the exact origin of Zinfandel, one of the most important red wine cultivars of California, has been a mystery to Americans ever since its arrival on the American continent in the early 1820s. And a mystery it remained, until Professor Meredith, with the help of her Croatian colleagues, decided to uncover the roots of this important variety and put an end to the controversy that intrigued California winemakers for decades.

My own interests in grape genetics would probably remain completely undiscovered had I not joined the Meredith team during my senior year of biochemistry studies at UC Davis in 1997. Although I had spent seven years at Davis, next door to the beautiful vineyards of Napa valley, I have to admit that my first encounters with California grapes occurred in the lab. This is the place where I became carried away by the problem-solving power of real, practical science.

It was pure luck that brought me, a Croatian native, to the laboratory exactly at the beginning of the "Zinquest". Motivated by my own Croatian heritage, I gladly accepted the invitation to participate in the fieldwork and, later on, research related to uncovering the origins of Zinfandel in Croatia. In the period from 1998-2002, I returned to the lab on several occasions to work on DNA profiling of the most interesting Croatian varieties, believed to be close relatives of Zinfandel. With the steadfast help of Gerald Dangl, the lab supervisor, and Summaira Riaz, a doctoral student at the time, I learned the steps involved in a real scientific experiment and had fun all along the way.

Shortly before the completion of the "Zinquest" and the discovery of nine Zinfandel vines (locally known as Crljenak kaštelanski) in Kaštel Novi, in 2001, I defended my Ph.D. thesis entitled "Investigation of relatedness between Zinfandel and autochthonous Croatian grape varieties (*Vitis vinifera* L.)" at the University of Science in Zagreb. For me, the doctoral thesis represented the closure of an experience that has been incredibly rewarding, both personally and scientifically.

This book is intended to present, in layman's terms, the ups and downs of the search for Zinfandel in Croatia from an "insider" perspective. For scientists, it should read as a "detective's story"; for historians, it is an account of Dalmatia's history reconstructed through "a vine trail"; for wine-lovers, it is a guide to the wine and vine resources of the Dalmatian coast; and for casual tourists, a picturesque portrayal of Croatia and its rare natural beauty.

Jasenka Piljac

ACKNOWLEDGEMENTS

Zinfandel: A Croatian-American Wine Story is a result of synchronized work of many people. For my interest in grapes and my first true love for science, I owe my gratitude to Professor Carole Meredith. For their valuable contributions and help with the publication of this book, I am most indebted to Professors Sibila and Vladimir Jelaska, and Želimir Bašić. For critical review of the book and useful suggestions from the perspective of a wine historian, a writer, and a Zinfandel expert, I am very thankful to Professor Charles Sullivan. For grammar and language corrections, and a final polish on the manuscript, I am grateful to Professor Mary Anne Saunders. I owe to Jagoda Bush many thanks for her shared enthusiasm and support in the past year. If it weren't for Jagoda's incentives, the book probably would never get published.

For continual unselfish support of my research work at "Ruđer Bošković" Institute, I thank professors Stjepan Marčelja, Nikola Ljubešić and Dražen Vikić-Topić.

For his motivation and spirit in the search for original Zinfandel and creation of a worldwide reputation for Croatia, I have to point out Mike Grgich, a unique personality in the winemaking world. And to all the scientists who participated in the "Zinquest" and to those less known before them, who realized the importance of preserving Croatia's viticultural treasure and ampelographic recordings, I dedicate this small contribution.

CHAPTER ONE

WINEGROWING IN CROATIA - THE LEGACY OF OUR FOREFATHERS

Tracing back the origins of the vine

The grapevine, *Vitis vinifera L.*, together with the olive, fig and the palm comprise the oldest group of fruit trees that have been extensively exploited in horticulture-oriented civilizations of the Old World. The stagnations and advancements of ancient peoples throughout history, from the Early Bronze Age until today, were marked by a continual cultivation of these crops.[1] The grapevine has been a devoted follower of many civilizations in their conquests, dispersions and, often, the one certain proof of their existence.

The predecessor of the cultivated grape, the wild vine, *Vitis sylvestris*, belongs to the same genus that comprises several dozen species. All members of the genus are perennial woody climbers that require extensive care and yearly pruning, for regulation of vegetation and fruit production.[1] In the past, botanists have regarded wild grapes as a separate species, due to their morphological similarity and relatedness. *Vitis sylvestris* is now considered to be the wild form of the domesticated grape. Therefore, *Sylvestris* vines have been classified as the subspecies within the *V. vinifera* species and are believed to be the original biological material from which the cultivated grape emerged. *Sylvestris* grapes seem to be indigenous to the mild climate of the Caspian belt and the southern coast of the Black Sea; however, they also abound in the vegetation belt starting from Turkey, across Greece, Croatia, Italy, France, Spain and even north-west Africa. The path of the cultivated vine seems to closely follow the wild grapes; both of these forms have adapted to the winegrowing regions along the Rhine and Danube rivers of central Europe. It is hard to draw a precise boundary between the cultivated grape and the wild vine forms due to the spontaneous cross fertilizations that took place between the two in the past.

Documented evidence of the first use of wine does not exist, although historical evidence points to Mesopotamia and the banks of the river Nile as the most probable home of the vine. Proof of this ancient tradition has been found in the form of grape seed remnants in villages dating from several thousand years B.C.[2] In approximately 1500 B.C., the cultivation of the grapevine was introduced to Greece and from there spread all over Europe.[3] Today, viticulture is most successful in relatively mild climates of western and central Europe, western Asia, and the temperate zones of the New World.

Besides the sweet and juicy fruit, the ancient peoples soon discovered another valuable grape product, its nourishing and enchanting wine. The antiquity of wine is indicated by the terminology which refers to this "drink of the Gods". In 1500 B.C., the Hittites referred to wine as *uiian* or *uianas*; the Greeks called it *woinos* or *ionos*, and in Latin and in other western languages it was referred to as *vinum*, or alternatively *vino*, *vin*, *wein*. The Semitic languages document the word *wayin* (later *yayin*) in Hebrew, *wayin* in Sabaean, Arabic, and Ethiopian.

Some of the best-preserved records of wine industry come from Egypt. Hieroglyphics indicate that Egyptians had a well-developed grape industry, with wine presses and preservation jugs. Remains of grapes and signs of wine production were found in Old Kingdom tombs. Among the earliest findings were small, carbonized pips belonging to the 1st Dynasty graves (2900-2799 B.C) in Abydos and Nagada. The iconography of the afterlife indicates that wine was one of the privileges of the world beyond. It is believed that, although the wine industry was well developed, it was not widespread, as the consumption of wine was restricted to priests and royalty. Wine then was probably of poor quality and must have been drunk soon after fermentation, before it turned into vinegar or got spoiled. The problem of the dilution of wine with water was noted as early as 1792-1750 B.C. in Mesopotamia, in the Code of Hammurabi.[2] Article 108 of the Code states that: "any person who purposely lowers the quality of wine shall be thrown into the water".

With the development of Greek civilization, the wine industry experienced significant improvement. Not only was wine considered an addition to the meal, rather, it became recognized as beneficial for medical purposes. Wine was a source of relief and comfort, as pictured in the *"Iliad"*, when Hector returns to Troy and his mother, Hecuba, gives him wine. Subsequently, the Greeks developed a wine trade and Greek colonies spread the culture of the vine as far west as Spain and as far east as the shores of the Black Sea.

The Romans inherited valuable knowledge about viticulture and winemaking from the Greeks when they settled in Italy in the 8th ct. B.C. However, the additional contribution of Romans to grape growing and wine production should not be diminished. Pliny[4] was the first to classify grapes and draw distinctions among varieties according to color, time of ripening, soil preferences, diseases and quality of wine. Historians speculate that, although they also used amphoras, the Romans introduced wooden cooperage – an important advance for the wine industry because it prevented air contact for a longer period of time. It is also believed that the Romans were very close to discovering pasteurization; they noticed that in warm conditions wine resists the undesirable changes causing spoilage. However, they were far from the discovery of major problem causers – microorganisms.

Even though wine was present in many ancient religions, Christianity has greatly aided the spread of viticulture through the use of wine during worship and the encouragement of grape growing on the premises of various monasteries.[5] The Christians allocate special meaning to viticulture in the context of historical events that shaped the foundations of their faith. According to the Bible, after the flood and Noah's salvation, God commanded him to mutiply and subjugate all living things on the earth. Noah first planted a vineyard and made wine:

"…and Noah began to be an husbandman, and he planted a vineyard; and he drank of the wine and was drunken." Genesis 9:20-21.

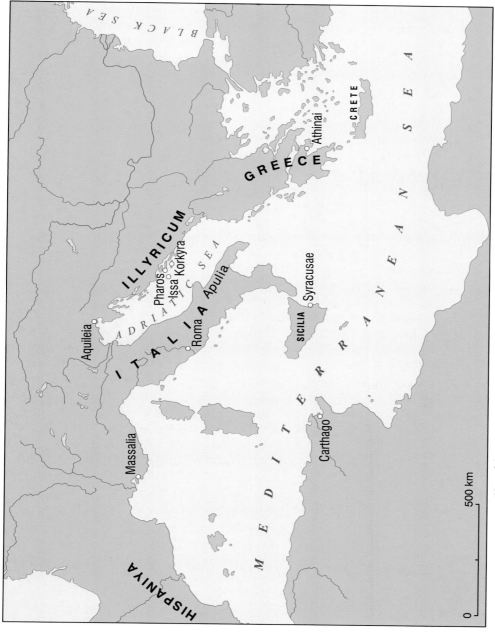

Map of the antic Mediteranean basin - the cradle of Vitis varieties (drawn by Tomislav Kaniški).

Christianity slowly spread throughout the Empire all the way until the Roman state granted formal recognition in 313 A.D. The acceptance of Christianity complemented the advancements in viticulture on the Balkan peninsula and winegrowing experienced a special boost during the Roman rule. Although the Roman Emperor Domitian (Titus Flavius Domitianus) forbade the planting of vineyards, soon after him, Emperor Probus annulled this decree and encouraged viticulture and trade in the regions of Rhine and Pannonia. This was the golden period for viticulture in Croatia and Slovenia.

Grape-growing tradition in Croatia

The first known people to colonize the shores of the Adriatic were the Phoenicians and the Greeks. And, although we can never be certain whether the earth hides even older secrets, archeological evidence and historical writings point to the presence of a wine industry on Croatian land back in the 11th ct. B.C.[6] Wine jugs belonging to the 6th century before Christ and meaningful names, such as that of a region on the eastern coast of Istria, (Kalavojna — in Greek — "good wine") provide additional proof of an ancient winemaking tradition on Croatian territory. In fact, it is believed that the first known inhabitants of present Croatia, the Illyrians, cultivated the vine even before the Greek and Roman times. Although historians and archeologists claim that ceramic pottery and wine cups of various dimensions found among the remains of Illyrian settlements in Ripač near the river Una, in Otok near Sinj and in Podvršje northeast from Zadar, cannot be taken as evidence of a developed wine industry, they certainly confirm that the Illyrians were familiar with the vine and its products. Since Illyrians were a large ethnic group, consisting of numerous tribes, many of which were unaware of each other's existence, it may be speculated that the vine was not evenly represented among them.

The invasion of Romans onto the Balkan peninsula began with the invasion of Istria and the surrender of the Illyrian tribe, the Histrions, in 179-178 B.C. Upon their arrival, the Romans discovered a well developed winegrowing industry, especially in Dalmatia (coastal Croatia) and on the Adriatic islands. Fortunately, the Romans considered agriculture the source of their power, and they encouraged its progress wherever the climate conditions were favorable. Numerous archaeological deposits from the Roman times confirm the presence of a widespread wine industry in this region. There are approximately 600 deposits along the Adriatic coast alone; the discovered utensils include amphoras, pitchers, dishes made of clay, metal and glass. The deposits found in Solin (*Salona*), once the capital of Dalmatia and the largest Roman settlement in present-day Croatia testify to the existence of an ancient grape growing tradition. A tombstone found in Poreč on the Istrian peninsula dating from 75 A.D. pictured two intertwining vines and a harvester picking the grapes. The presence of Latin terminology in Croatian viticulture and enology persists to this day in the following terms: vino(*vinum*)-wine, bačva(*bicus*)-barrel, bokal(*pocalum*)-jug, kosijer(*cossero*)-knife, krba(*corbula*)-basket.[6]

Reconstruction of an Illyrian merchant ship, liburna (Liburna navis)
(Source: Želimir Bašić - Dalmatinska vina kroz stoljeća, Šibenik 2001).

During the reign of Emperor Claudius (41-54), entire hills of Mon Claudius (today Moslavačka gora) were under vineyards. However, with the change of rule, and the coming of Emperor Domitianus (81-96) in power, viticulture suffered great losses due to new laws forbidding winegrowing in the conquered provinces. Namely, the Romans, afraid of the fierce competition with the Italian industry, decided to restrict viticulture and especially the production of quality wine. Croatian winegrowing started to decline until Emperor Probus (276-282) again encouraged the spread of vine cultivation, especially in Pannonia.

The improvements made in viticulture by the Romans, especially in Pannonian and Dalmatian winegrowing, were devastated during invasions by the Slavic tribes.

A Roman tombstone with vine motifs unearthed in Isakovci, dating back to the 2nd ct. A. D. (Source: Želimir Bašić - Dalmatinska vina kroz stoljeća, Šibenik 2001).

Only the remote Dalmatian islands and parts of Istria resisted invasions and plunder. There are few sources that give information about winegrowing in the territory of Croatia after the fall of the Roman Empire. However, according to later records,[6,7] it remains certain that the newcomers soon adopted the old Roman traditions and turned to agriculture for prosperity.

The first certain proof of Croats' ownership of vineyards comes from a decree by which the Croatian Prince Mutimir, in 892, donated "ploughlands and vineyards, pastures and woods" to the Split church.[7] All Dalmatian municipalities grew vines, and regulations controlling wine prices were adopted: a law of the two city councils of Dubrovnik from 1424 prohibited the raising of wine prices, except for the wine Malvasia, which could be sold at the price set by the winemaker. In fact, this is the first document of a Croatian wine sold under the variety name.[8] The legend says that upon tasting Malvasia, the Pope removed the curse previously called on the citizens of Dubrovnik. Additional regulations, such as the statute of Hvar from 1331, or Korčula from 1407, prohibited wine import, thus protecting Croatian winemaking. Getting drunk was also considered shameful and punishable, while sobriety and hard work on the scarce soil were greatly encouraged. The harvesters found a middle ground in this dilemma by introducing the famous five-century-old Dalmatian drink, *bevanda* – part wine and part water.

The development of Dalmatian ports served the purpose of promoting wine trade between Dalmatia and its neighbors. The new laws of Venice in 1402, permitted wine import and threatened the economic stability of Dalmatian communities which, until then, had a monopoly in this branch of agriculture. At the same time, the monasteries (specifically the monasteries of Zagreb, Čazma, Požeški kaptoli, Kutjevo, Rudine) became more and more important in promoting viticulture. The cult and symbolism of wine, and its importance in mass celebrations, spread into the everyday lives of Christians (Catholics).

In the second half of the 15th century, winegrowing was again threatened by the frequent invasions of the Turks, which were marked by devastation, slaughter and destruction. The hardest hit territories were left without the manpower needed to work the fields. Srijem and middle Slavonia suffered an even worse fate and finally surrendered to the Turks in 1536 and 1537, respectively. Dalmatian vineyards were not spared either. In the long period of Venetian-Turkish wars, the worst damage was done to the vineyards around Šibenik, Trogir and Split, where earlier documents recorded the existence of more than 7000 vineyards.[6,7] Although the Venetians eventually succumbed to the fierce attacks of Turkish invaders, their influence on viticultural practices in coastal Croatia was felt for years to come.

Concerns over Ottoman expansion, and the destruction that went along with it, led the Croatian Assembly to seek the help of the Habsburgs, under the rule of Archduke Ferdinand. Owing to the persistent resistance of the Habsburgs, by the 18th century, most of Croatian territory was freed from the Turks, and Croatia regained its domestic autonomy.

As a member province of the Austro-Hungarian Empire, from 1867 until the end of World War I, Croatia was still under the strong influence of the Empire. Following World War I and the demise of the joint Austro-Hungarian rule, Croatia was, against the will of its people, included in the Kingdom of Serbs, Croats, and Slovenes, which became Yugoslavia in 1929. Under the communist leadership of Marshall Tito, the Croats, Serbs, Slovenes, and ethnic minorities were united in the new state – the Federal Socialist Republic of Yugoslavia. An ending to the Yugoslav federation was foreseen with the death of Tito and the fall of communism throughout eastern Europe. Croatia declared independence from Yugoslavia in 1991, and under the wartime leadership (the war for independence lasted from 1991-1995) of its first president, Franjo Tuđman, emerged as a sovereign young democracy in the heart of Europe.

After experiencing numerous rises and falls, in the midst of historical turmoil and changes, Croatian viticulture went through three main restoration periods.[5] The first restoration of winegrowing in Croatia began in the last decade of the 17th century, when new settlers with no experience in viticulture moved to Slavonia. They refreshed the Slavonian wine repertoire by introducing new varieties such as Burgundy, Traminac, Rhine Riesling, Zeleni Silvanac etc. Winegrowing once again spread throughout Croatia and was accompanied by a general increase in wine consumption. Viticulture even spread to those areas which had not been under the Turks; one regulation from 1780 indicates that all villages of the Jastrebarsko estate had their own vineyards, a total of 200.

As a consequence of the devastating effects of phylloxera, most European vineyards have been replanted on resistant American rootstocks; at the end of the 19th century. This second restoration was difficult because Croatia was going through social and economic changes where viticulture and winemaking lost their priority in Croatian agriculture. In an effort to prevent further loss, new varieties were introduced and the old ones grafted onto resistant American rootstocks.

The third restoration period that followed the Second World War emphasized the development of clonal selection, crosses and employment of industrial machinery in viticulture. With advancements seen in agrotechnology, ampelography, and mechanization, as well as continued introduction of new varieties, Croatian viticulture went through a revival, and winegrowing became an increasingly important branch of the economy. Today, the well-known wineries in Imotski, Benkovac, Šibenik, Stari Grad (Hvar) and Umag are a result of the work of experts, and fertile Croatian soil.

Croatia's viticultural treasure

Since the early beginnings until today, an enormous number of grape varieties have been grown in this geographic region. Certain sources cite that, prior to the arrival of phylloxera at the end of the 19th century some 400 cultivars were grown in Croatia[9], while it was possible to find several hundred cultivars in Dalmatia alone.[10] Most of these cultivars were considered to be autochthonous – long native to Croatia and long associated with this viticulture region. Besides favorable climatic conditions, a rich past and good connections with the other countries where grapes were grown had a strong influence on the cultivar number. It is probable that some of them developed in this area, and others were introduced a long time ago. Unfortunately, many cultivars were lost at the beginning of this century in vineyard destruction caused by new fungal pathogens and pests (peronospera, oidium, plasmopara, uncinula, phylloxera) as well as by the demands of modern production. Although the total cultivar number today is significantly smaller, and although certain genotypes have been lost beyond rescue, a large number of varieties considered to be autochthonous still remain.

The official Croatian cultivar list consists of 143 genotypes, both introduced and "domestic".[11] The 83 cultivars out of the mentioned 143 are considered to be long native, while an additional 60 extremely rare genotypes remain underutilized. Such diversity of genotypes within a country as small as Croatia (56,542 km^2) is probably due to varying climate conditions, from a cold and wet continental to a dry and hot Mediterranean climate. Despite the long history of vegetative propagation and possible introduction, the Croatian *Vitis vinifera* gene pool is rather independent and well established. Recent research results[12] emphasize the uniqueness of Croatian cultivars and reveal a substantial level of variation within the Croatian population of *Vitis vinifera* L. In fact, according to scientific parameters, only the Italian grapevine population exhibits higher diversity. Thus, it is of great interest to stress the importance of preservation of the valuable genetic resources represented by these cultivars. In an effort to ensure the preservation of autochthonous Croatian cultivars and prevent further extinction of unique genotypes, a collection has recently been established at the experimental vineyard station "Jazbina", at the Faculty of Agriculture in Zagreb.

Native grapevine cultivars grown in Croatia included in the official list of grape cultivars issued by the Ministry of Agriculture (article no. 12/94 and article no. 96/96). The cultivars marked with an ^{s} consist of two or more subcultivars (types).

Code	Cultivar name	Color of berries	Main growing area	Subregion
1	Babić crni^s	B	Dalmatia middle and north	SJD, SD
2	Blatina crna	B	Dalmatian hinterland	DZ
3	Bogdanuša bijela^s	W	Island of Hvar	SJD
4	Botun bijeli	W	Dalmatian hinterland	DZ
5	Brajda crna^s	B	Dalmatia north	SD, HP
6	Brajdenica	W	Istria	IS
7	Bratkovina bijela	W	Dalmatia south	SJD
8	Cetinka bijela	W	Islands of Korčula, Hvar	SJD
9	Crljenak crni	N	Island of Vis	SJD
10	Debit bijeli^s	W	Dalmatia north	SD
11	Diseća belina	W	Croatia north-west	PB, ZM
12	Dobričić	B	Middle Dalmatia	SJD
13	Drnekuša^s	B	Island of Hvar	SJD
14	Dugovrst	W	Dalmatia north	SD
15	Galica crna	B	Dalmatia north	SD
16	Gegić bijeli^s	W	Island of Pag	HP
17	Grk bijeli	W	Island of Korčula	SJD
18	Gustopupica crna	B	Dalmatia north	SD, SJD
19	Hrvatica crna	B	Istria	IS
20	Kadarun crni	B	Dalmatia south	SJD
21	Komoštrica bijela	W	Dalmatia middle and south	SJD
22	Kraljevina crvena^s	W, R	Croatia north-west	PB, ZM
23	Kujundžuša bijela	W	Dalmatian hinterland	DZ
24	Kurtelaška bijela	W	Middle Dalmatia	SJD
25	Lasina crna^s	B	Dalmatia north	SD
26	Malvasia dubrovačka biijela^s	W	Dalmatia south	SJD
27	Malvazija istarska	W	Istria	IS
28	Maraština^s	W	Dalmatia	HP, SD, SJD, DZ
29	Medna bijela	W	Middle Dalmatia	SJD
30	Mekuja bijela	W	Island of Hvar	SJD
31	Mladinka	W	Middle Dalmatia	SJD
32	Moslavac bijeli	W	Croatia north-west	MS, PB, ZM
33	Muškat istarski	W	Istria	IS
34	Ninčuša	B	Middle Dalmatia	SJD
35	Okatac crni	B	Middle Dalmatia	SJD
36	Okatica bijela	W	Dalmatian hinterland	DZ
37	Opačevina	W	Istria	IS, HP
38	Pagadebit bijeli	W	Island of Korčula	SJD
39	Palaruša bijela	W	Middle Dalmatia	SJD
40	Plavac mali^s	B	Dalmatia middle and south	SJD
41	Plavac veliki	B	Dalmatia middle and south	SJD
42	Plavec žuti	W	Croatia north-west	PB, ZM
43	Plavina crna^s	B	Dalmatia	HP, SD, SJD, DZ
44	Pleškunača	W	Island of Susak	HP
45	Pošip bijeli^s	W	Island of Korčula	SJD

Code	Cultivar name	Color of berries	Main growing area	Subregion
46	Prč bijeli[s]	W	Island of Hvar	SJD
47	Ranfol bijeli	W	Croatia north-west	PB, ZM
48	Ranina bijela	W	Croatia north-west	PB
49	Rudežuša	B	Dalmatian hinterland	DZ
50	Ruževina bijela	W	Dalmatia north	SD
51	Ružica crvena	B	Dalmatia north	SD
52	Sušćan	B	Island of Susak	HP
53	Sušić	B	Island of Susak	HP
54	Šipelj	W	Croatia north-west	PB
55	Škrlet bijeli[s]	W	Moslavina	MS
56	Trbljan bijeli[s]	W	Dalmatia	SD, SJD
57	Trnjak crni	B	Dalmatian hinterland	DZ
58	Trojišćina crvena	R	Island of Susak	HP
59	Vugava bijela	W	Island of Vis	SJD
60	Zadarka crna	B	Dalmatia middle and north	SD, SJD
61	Zelenika bijela	W	Croatia north-west	ZM, PB
62	Zlatarica bijela	W	Dalmatia middle and south	SJD
63	Žilavka	W	Dalmatian hinterland	DZ
64	Žlahtina bijela	W	Island of Krk	HP

Rare native grapevine cultivars grown in very limited areas in Croatia which are collected at very specific sites (mostly islands and very specific microclimate), primarily selected according to their local names and local importance. There are very limited ampelographic and other data about them and there is certain likelihood that some of them are just synonyms of some known cultivars.

Code	Cultivar name	Color of berries	Main growing area	Subregion
65	Bratkovina crna	B	Island of Korčula	SJD
66	Pošip crni	B	Island of Korčula	SJD
67	Topol bijeli	W	Island of Pag	HP
68	Muškatel	W	Island of Pag	HP
69	Petovka	W	Island of Pag	HP
70	Cipar rumeni	R	Island of Pag	HP
71	Silbijanac	W	Island of Pag	HP
72	Krstičevica bijela	W	Island of Hvar	SJD
73	Gargičevica bijela	W	Island of Hvar	SJD
74	Ruža bijela	W	Island of Hvar	SJD
75	Mirkovača	R	Croatia north-west	PB, PK
76	Ovčiji repak	W	Croatia north-west	PB, PK
77	Barjanka bijela	W	Island of Korčula	SJD
78	Rumenka	R	Island of Korčula	SJD
79	Klešćec bijeli	W	Croatia north-west	PB
80	Divjaka bijela	W	Dalmatia middle and south	SJD
81	Lipolist	W	Dalmatia middle and south	SJD
82	Lelekuša crna	B	Island of Hvar	SJD
83	Muškat ruža crni	B	Dalmatia middle and south	SJD

B - black, W - white, R - rosé

According to the national statistics from 1999 (National Institute of Statistics[13]), the area under vineyards in Croatia amounts to 59,000 ha, which yields an average of 390,000 tons of grapes and 2,100,000 hl of wine per year. These figures are questionable because Croatia has just begun the development of its wine land register, which will include all the relevant data about the vineyards in the state, independent of their purpose, area and the nature of maintenance. Viticultural regions of Croatia are divided into seven continental growing subregions and five coastal growing subregions. The production of grapes and wine is a significant activity for many Croatians – there are indications that over 10% of the population earns a living from vine products.

The significance of the extent of viticultural activities currently underway in Croatia, and the position of Croatia as an important contributor to European wine production, have recently been recognized by the international office for vine and wine (OIV – Office International de la vigne et du vin), founded in Paris in 1924 (oral communication with the Croatian national institute for viticulture and enology). Croatia was granted equal membership in the OIV in October of 2001, with all the accompanying duties and responsibilities. As a member-state of the OIV, Croatia will have the opportunity to voice its concerns not only among the local winegrowing community, but at an international level as well. This important recognition will enable Croatian winegrowers, experts and scientists to participate in the shaping of international wine policy, as well as work on promoting their own ideas and interests related to viticulture and winemaking.

Wine in Croatian folk customs

It is believed that wine follows a man from his cradle to the grave. When a child is born, it is customary to make a toast to its health and happiness. In many religions, in the funeral mass, wine is drunk as a symbol of reunion with Christ in the afterlife. In Croatian folk customs, wine has found its place as a symbol of good will, health and prosperity. Croatian oral and literary opera, especially sayings and proverbs, are rich with references to wine. Wine is an unavoidable item and, often, the central theme of social gatherings, family and holiday celebrations. Croatian folk customs and beliefs related to wine have their roots in legends and religious traditions. The completion of seasonal activities in vineyards, often connected with church holidays, has traditionally been the occasion for gathering of vintners in homes and vineyards.

The life in vineyards is awakened after a winter rest, on January 22 – St. Vincent's day (Sv. Vinko). St. Vincent was born in Huesca, Spain, and was a renowned Christian in Saragoza. He lost his life in the last persecution of Christians in the 4th ct. A.D., during the reign of Emperor Diocletian, and was later proclaimed the saint-protector of vineyards. St. Vincent's day in Croatia is marked by the blessing of vines with old wine and holy water, and the cutting off of the young shoots. According to one old custom, the shoots are placed in water in a warm room until they grow leaves, buds and clusters. The vines are "evaluated" according to the

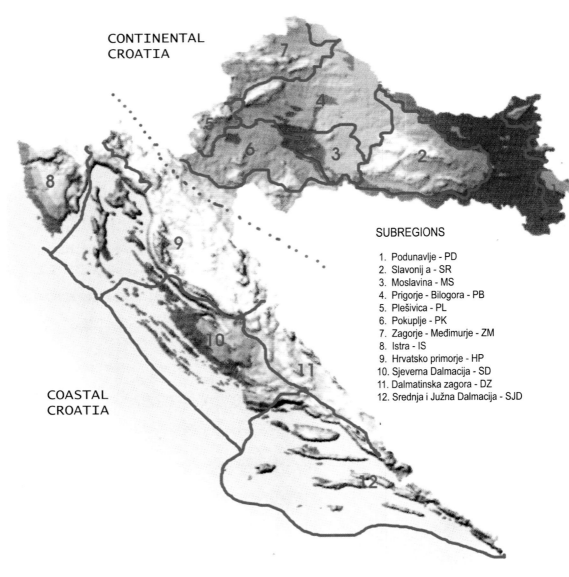

Viticultural regions of Croatia (Source: Several authors - Hrvatska vina i vinari, Zagreb 2002).

number of clusters, and the next year's yield predicted. From one district to another, the ritual somewhat varies. In certain places, the blessing of vines is accompanied by tambura music – a stringed instrument widely played in Croatia. Often, the vintner performing the ritual directs words of hope to his vineyard by saying: "I give you old wine, in exchange for new grapes and must", or "Dear God, please look after my vines".[14]

Springtime is a period of extensive activities in vineyards and there is little time left for celebration. However, in Slavonia - the far eastern province of Croatia, March 19th marks St. Joseph's day and the day when the grafting of vines begins. According to the legend, Joseph was predestined to be Virgin Mary's fiancée. As a sign of God's choice, when laid upon the altar in the Temple, Joseph's dry cane miraculously grew leaves. In associating this happening with the vine's biological cycle, the winegrowers are hoping to encourage the budding of vines in early spring.

The summer is a period when vineyard duties are not so demanding. However, the grapes need to be protected from diseases, pests and environmental hazards. This is why, during the summer months, it is customary to light candles in vineyards to prevent hail, set up wooden rattles as a warning for the wind, and scarecrows as prevention against birds and other pests. In summertime, thorough preparations are performed for the upcoming harvest activities, which start in mid-September and end in late October.

Finally, the arrival of autumn brings about the harvest, which in most winegrowing regions begins on St. Mihovil's day, the 29th of September. The harvest is a family celebration that follows after extensive preparations in vineyards. Family, close and distant relatives and friends usually participate in the festivities following hard

Harvest procession in Dalmatia (photo by Boris Kragić).

Harvest activities in Dalmatian vineyards (photo by Boris Kragić).

work in the vineyards. Historians[15] have recorded that in the past, the vineyards in Croatia were especially alive with song in harvest time. In his description of the harvest in Slavonia, recorded in 1857, Mijat Stojanović noted: "The vineyards become very lively during the harvest…the entire neighborhood bursts in song, the bells ring, the musicians play and the folk make jokes". Since this is a very busy part of the year for the vintners, there is little time left for ceremonies; however, the traditional good-will rituals are always performed. Upon completing the harvest, it is customary to build a bouquet of healthy grapes, vine leaves and canes and take it home to the vineyard owner, as a symbol of good fortune. The bouquet is usually preserved until Christmas time. The meals prepared for the vintners during harvest are often rich and the women make sure that their tables are abundant with sweet delicacies and old wine, while the host prepares the main roast.

Traditionally, the end of harvest in Dalmatia is celebrated on the first Saturday after the completion of all harvest-related activities in the region. This celebration, also known as Sabatina after the Italian word "sabato", is marked either by tears or laughter, depending on that year's yield. On this day, the farmers traditionally assemble in order to discuss the results of the harvest, exchange their experiences and suggestions, and taste each other's wine. Sabatinas organized on the islands often turn into very lively social happenings and have become major wine-related festivals in Dalmatia.

After the harvest, with the supplies of last year's wine mostly depleted, the winemakers eagerly await their new young wine. The key date in celebration of young wines is St. Martin's day, November 11th. As is the case with most of the other wine saints, St. Martin was not in the past related to wine or Dionysian festivities. He was born around 316 in Hungary.[14] According to the legend, he came from a gentile family but his own beliefs and faith in Christianity were very strong. He became a bishop in 371, especially devoted to the peasants and improving the life in small villages. He died on November 8th, and was buried three days later, on November 11th. Throughout Croatia, this holiday is celebrated and it is ordinarily believed that on St. Martin's day the must ripens and turns to young wine. The blessing of young wine the day before Martinje is in the center of many festivities held throughout Croatia. It is customary for small groups of men to assemble in the vineyards and perform a specific ritual, very much similar to a religious rite, believed to be the heritage of St. Martin himself. Often the host - the owner of the vineyard - allocates a "bishop", the "ministrants", and a "godfather", who, dressed in the appropriate festive clothing, gather around the table where food, salt, young wine and a candle have been placed. The host then sprinkles the salt into the wine jug, saying: "I bless you must, and turn you into wine, in the name of the one who prepared you from the earth and in the name of St. Martin and his spirit that makes you fizzy..."[15] The jug full of wine then circles the table so that all the guests may taste it, ending the ceremony. This custom is still common today, with small variations depending on the winegrowing region and the locals. The festivities of St. Martin are often retold and remembered until the next November 11th, when new young wine calls for its blessing.

Superstitions related to wine and the harvest are also common in Croatia.[15] Once upon a time, the grape growers believed that the grapes should not be harvested in rainy weather, because the wine made from such grapes would be weaker and of lesser quality. A common saying summarizes this: "September rains are worth gold in the fields, but mean poison to the vineyards". Numerous other well-known Croatian proverbs and sayings, often offering advice to the winegrowers, are a consequence of cherishing the long and laborious tradition throughout the countries: "Few vines, many grapes – many vines, few grapes", "If you want a fertile vine, hoe up before September-time", "To renew old vineyards and vines, prune before the leaf withers, while it still shines", "Beware - head pruning might leave you crying".

Traditionally, it was believed that the weather conditions on certain dates or periods in the year can predict the quality of next year's wine. For example, if January is rainy and wet, then the wine at St. Martin's time will be scarce. If February is windy and cold, then the year will be fruitful. A dry March, a wet April and a cold May promise rain in the summer and good wine in the fall. A dry June usually means that the wine will be good, and a hot August ensures "hot" (strong) wines. If August 15th, a major Catholic holiday – the Assumption of the Virgin Mary, is a bright and sunny day, that year's wine will be good. If St. Martin's day is covered with snow, the year will be fruitful, but if trees set new flowers on that day, then the next year will be poor.

The vine and wine have in the past been attributed healing powers, as witnessed by numerous folk customs still common in Croatia today. Wine was used as an antiseptic, fever reducer and in treatments against hair loss. Honey was often spread across the vine leaf and, when applied to sores and wounds, promised fast relief. Blessed wine was believed to possess special properties; without it, the holy rituals had little meaning.

Wine has also played an important role as a source of inspiration to many contemporary Croatian artists, writers and musicians.[16] The painter Zoe Borelli Vranski-Alačević has captured the moment of child baptism with wine, an old Dalmatian custom performed to the male firstborn in the family, intended to protect him and ensure his happiness. Her painting from 1939 is kept in the collection of the Etnographic museum in Split, as well as her later work from 1940, also portraying a wine-related custom. In the not-so-distant past, before major church holidays in small towns and villages of Dalmatia, it was customary to place a wooden barrel with wine on top of the water wells located in front of house entrances. As a sign of good spirit, the host offered a special kind of dry pastry called "el buzola duro" or "desop" and his wine to the neighbors and passer-bys.

To this day, the sensational applications and the uniting power of wine have been unmatched. The use of wine and its symbolism in folk customs testify to the importance of this drink in Croatian tradition and cultural heritage. It is undoubtable that "the wine has integrated itself in our history in such a way that it may never disappear in it", as Professor V. Belaj of the University of Zagreb once declared.[14]

Significant periods in Dalmatia's viticultural history

The antic period

As mentioned earlier, the first certain evidence confirming the use of vine in Dalmatia dates back to the antic period, although the precise year of introduction of this crop is unknown. It is possible that the vine migrated from the southern shores of the Mediterranean sea, or that it traveled along the northern route of the Danube river. Its exact trail that lead to Croatia is uncertain, although some even argue that the cradle of this important crop is here, where it underwent a genetic transformation from the wild *Sylvestris* predecessor whose traces are easily found along the Adriatic coast today.

The long-held viewpoint ascribing the development of viticulture in this region only to the Greeks has been proven false by the historians. Archeological evidence in Otok near the town of Sinj, dating back to the Bronze age (1800 B.C.),[17] indicates that a developed winegrowing Illyrian civilization existed long before the Greeks. However, we do owe to the Greeks the oldest existing document confirming the cultivation of the vine on Croatian land, the work of the Greek writer Athenaious entitled "The feast of scholars" (2nd ct. B.C). In this document, Athenaious quoted the words of his predecessor Agarthchides (2300 B.C.) praising Dalmatian wine: "On the island of Issa (Vis) in the Adriatic, people make wine better than all other wines compared with".[6]

ἐν δὲ Ἴσσῃ τῇ κατὰ τὸν Ἀδρίαν νήσῳ Ἀγαθαρχίδης φησὶν οἶνον γίνεσθαι ὃν πᾶσι συγκρινόμενον καλλίω εὑρίσκεσθαι.

Agarthchides came from Knidos – a city located in the Aegean sea, near the island of Rhodes. He was a writer by profession and his most noted work was "Red sea". Unfortunately, only fragments of Agarthchides' writings have been preserved and later presented by other authors as quotes. According to the legend, Agarthchides was Ptolomeous' teacher and on his numerous journeys throughout the antic world, he had the opportunity to taste a myriad of wines. Thus, his impression of Issean wine may be taken as knowledgeable and objective.

Greek influence

The Greeks established their city-states, *polis*, on the islands of Korčula (Korkyra, 6th ct. B.C.), Hvar (Pharos, 385 B.C.) and Vis (Issa, 397 B.C.) across the sea from Split. Precisely on these islands is where the cult of wine had its most fervent followers. The city-state of Issa was the strongest Greek colony, with its own statute, local administration and 38 different kinds of coined money. The six-gram coins had a grape relief imprinted on the front and an amphora on the backside. Smaller coins showed a grape vessel with two handles on the front and a grape cluster on the back. Among others, coins carrying the image of Dionysius and the goddess Athena on one side and grapes on the other were also discovered. The entire

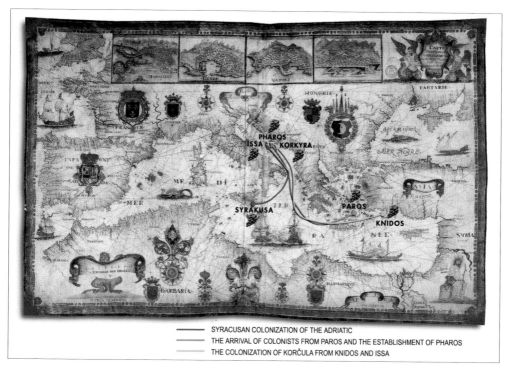

The directions of Greek colonization of Dalmatia.
It is believed that the Greeks brought some of their cultivars from Paros, Knidos and Syrakusa to the Dalmatian islands of Issa, Pharos and Korkyra (Source: Želimir Bašić - Dalmatinska vina kroz stoljeća, Šibenik 2001).

Lumbardian psefismus: a detail from the original stone script.
(Source: Želimir Bašić - Dalmatinska vina kroz stoljeća, Šibenik 2001).

series of coins probably originates from the period between 225-200 B.C. In Lumbarda, on the island of Korčula, archeologists uncovered a stone plaque, the *Lumbardian psefismus*,[17] a document dating back to 450-500 B.C. witnessing the existence of a Greek colony and its early regulations governing the distribution of winegrowing regions on the island. An excerpt from the *psefismus* clearly states:

"..the founders of this settlement have made an agreement, and the people have agreed, that those who settle first and build a fort for this city shall take a piece of land for a house within the city walls...and shall have priority in choosing the land for vineyards across the area of three pletars, outside the city walls..." (one pletar equals about 9556 square feet ≈ 867 square meters)*

This unique helenistic stone plaque precisely determined the amount of land allocated to each member of the colony for a house (1.5 pletar) and for a vineyard (3 pletars), in total 4.5 pletars. The continuity of winegrowing in Lumbarda is symbolically represented by the remains of the city walls of Korkyra that today serve as boundaries between vineyard properties near the church of St. Cross.

Ancient coin (front and back) dating back to the Greek period, found in Lumbarda on the island of Korčula
(Source: Želimir Bašić - Dalmatinska vina kroz stoljeća, Šibenik 2001).

---

*   Free translation of the Archeological museum in Zagreb where the original is kept

Some historians have argued that one of the most valued cultivars, Grk bijeli, was brought to the island by the Greeks. Others believe that this variety, well-known for excellent dessert wines and the highest quality white wines with protected geographic origin (Lumbarda appellation), is native to the island and that its name was derived from Croatian adjective "grk" referring to the bitter taste of its grapes. One thing is certain, from the Greek period until today, the vine continued to be the most represented agricultural crop on Korčula.

Greek mythology is rich with references to wine; the Greeks worshiped Dionysus – the god of wine and, in his honor, organized luxurious festivities and sang special songs called the *ditiramb*. Dionysus symbolizes fertility, love, pleasure and

Greek wine jug (oinohoe) found on the island of Vis, originating from the 4-2 ct. B.C. (Source: Želimir Bašić - Dalmatinska vina kroz stoljeća, Šibenik 2001).

elation, emotions most often associated with the narcotic effects of wine. It is believed that the Greeks initially worshiped the Egyptian god of sun and fertility, Osiris, and later renamed him Dionysus (in Greek, Dionisos). According to the legend, Dionysus is the son of the supreme Greek god, Zeus, and his lover Semele. Upon learning of their affair and Semela's pregnancy, Zeus' jealous wife Hera prompted Zeus to show his power and might by sending down on earth flashes of lightning and roles of thunder. Unaware of her secret plan, Zeus obliged. As a consequence, his pregnant lover Semela prematurely gave birth to Dionysus. When

Wine jug with Dionysian motifs and an enlargement of figures associated with the god of wine (Source: Želimir Bašić - Dalmatinska vina kroz stoljeća, Šibenik 2001).

Zeus discovered Hera's evil intentions, he wrapped thick ivy around the newborn and saved his son from death.

Most sources point to Theba as Dionysus' place of birth, from which place Zeus sent his son to the valley beneath the hill Nis. According to the myth, in order to protect Dionysus from Hera, the nymphs hid him in a cave the entrance to which was covered with vine leaves. In his hideout, Dionysus learned the trade of winemaking and continued to spread the knowledge of the vine to the shepherds. The news of Dionysus and his omnipotent drink soon spread everywhere around, to the joy of his father Zeus, who proclaimed him "the god of wine and viticulture".[18]

Dionysus was the favorite deity of widely differing social classes, because of his beauty, strength and associations with prosperity and fertility. Along with their experience and knowledge of the vine, the Greeks brought to the Adriatic coast the cult of Dionysus and Priapus – the lesser-known god-protector of gardens and vineyards. According to Greek mythology, Priapus is the son of Dionysus and Aphrodite – the goddess of love and beauty. Priapus' torso is often shown naked with rich decorations portraying grapes and vine leaves. A well-preserved stone statute of Priapus with relief grape decorations dating from the late Roman period is kept in the Archeological museum in Split.

A stone statue of the Greek god Priapus, the god-protector of vineyards, kept in the Archeological museum in Split (Source: Želimir Bašić - Dalmatinska vina kroz stoljeća, Šibenik 2001).

Roman influence

The Roman civilization accepted and continued the viticultural tradition of its predecessors. Archeological findings, stone wall decorations, amphoras and wine jugs, ceramics and pottery characteristic for this period, can still be found today on the Pelješac peninsula and the surrounding islands. On Korčula, the Romans cultivated the vine across the entire island. The remains of the *villae rusticae*,[19] usually built in the centers of vineyard properties, are still found today in Lumbarda, Blato, and Vela Luka. Pelješac, along with the Istrian peninsula in northwestern Croatia, are still the most significant wine producing regions of Croatia.

In A.D. 92, the Roman Emperor Domitian enforced a decree whereby half of the vineyards in France were uprooted by the legions in order to eliminate the competition with Italian wines. At that time, Italian viticulture suffered seriously and the winegrowers felt threatened because high yields were obtained in French winegrowing regions. The vineyards were cultivated secretly for two centuries in France until the third century A.D., when Emperor Probus (276-282) annulled the infamous decree. Aware of the winegrowing potential of his provinces, Emperor Probus encouraged the spread of viticulture from continental Croatia to the islands and coastal Dalmatia, as well as trade between the Roman provinces. The legend says that Probus planted the first vineyards on the hills of Mons Alma and Mons Aureus.[7] From then on, viticulture prospered not only on the Adriatic coast but also in the inland provinces, such as Srijem, Baranja, Požeška kotlina, Moslavačka Gora, Hrvatsko Zagorje and Varaždin.

Historical documents indicate that among Probus' numerous trade shipments to the Gauls, grapevine cuttings found their place as well. Heunisch Weiss, a variety once widely grown in Europe and, ac-

Roman amphoras found in the depths of the Adriatic, off the coast of Dalmatia, preserved in the Archeological museum in Split (photo by Jasenka Piljac).

cording to Goethe's "Handbuch der Ampelograpie", native to Croatia, was probably one of Probus' gifts to the Gauls. This variety, definitely not indigenous to northeastern France, had been extensively grown there under the name "Gouais blanc" in the Middle Ages. The name "Gouais" derives from the French "gou", a derisory term. Since it was considered to be a mediocre variety, the cultivation of "Gouais blanc" was discouraged and this cultivar was eventually banned from French vineyards.

However, this old Dalmatian variety has definitely left its traces on French winemaking. In a scientific paper published in 1999,[21] genetic investigations conducted by Professor Carole Meredith of the University of California at Davis proved that the world-famous varieties, Chardonnay, Gamay Noir, Aligoté, Melon and 12 other varieties of northeastern France are the progeny of separate crossing events between Gouais Blanc and Pinot Noir. Pinot Noir is a highly acclaimed French variety; some argue that it was first referred to by the Roman agricultural writer Columella and that it was probably present in the Burgundy region ever since the Roman conquest.* Therefore, the surprise in this finding was the other

---

* As indicated by P. Viala and V. Vermorel in "Ampelographie", vol. I.-VII, Masson, Paris, 1901-1910.

lesser quality parent, Gouais Blanc, especially since Chardonnay wines and the other wines of Burgundy and Champagne regions have been known as the world's best for centuries. Scientific studies, though, have shown that highly acclaimed varieties may have humble and unexplored origins, which might be as interesting as their world-class reputations. The genetic potential of other autochthonous Dalmatian varieties, and the Zinfandel story that follows in the next chapter, may be of even greater interest to winemakers in the future.

The medieval period

The medieval period, often considered to be the "dark" period in terms of historical advancement, marks the continuation of wine production along the southeastern shores of the Adriatic. According to notary records and documents referring to medieval communities, wine was considered to be a food item in Dalmatian towns, and the local authorities encouraged its production and marketing. They even guided the distribution of wine products among all levels of society, pointing to the formation of early wine politics in the medieval period. All medieval communes regulated wine trade by statutes, one of the oldest of which is is the Zadar statute – Statuto di Zara (early 14 ct.), along with the Hvar (1331) and Korčula statutes (1407).[21] Viticulture education was encouraged, as it ensured better production and higher income in vine-dependent regions. The first books and "how-to" manuals (*cathechismus*)[22] became common during this period and were mostly used by traveling teachers whose goal was to spread the knowledge about this important craft. It seems as though the importance of wine in the medieval period had risen to the point where living without it meant degradation of the society – wine had to find its place at the table of each member of the community.

The importance of wine is easily noted in the Dubrovnik statute (medieval Dubrovnik had the elements of an aristocratic Republic, as well as a city-state, and special laws and regulations applied to this region), according to which illegal wine imports inside the city walls were punished by distribution of this valuable product to the poor.[23] The Dubrovnik Republic carefully regulated its wine imports and exports; of-

Wine trade in medieval ports - drawing by Antun Zupa (Source: Želimir Bašić - Dalmatinska vina kroz stoljeća, Šibenik 2001).

ten, high taxes were imposed on products coming from other medieval communes in order to protect the local wine production. At times, when the import of wine inside the city walls was completely forbidden, the mere thought of impartation or a question showing such an interest was severely punished. According to a regulation[23] from 1360, such a breach of conduct among the members of the Great Council of Dubrovnik was punished by 100 perpers (15 grams of silver). In cases of poor harvest, the taxes on wine imports were removed, and in times of wine crises, the funds allocated for grain were instead applied towards wine purchases. According to certain sources,[24] the annual production of wine in Dubrovnik during 1360s amounted to approximately two million liters, which averaged about 517.5 liters of wine per resident. Philosophical and didactic literature from the medieval period also points to extensive wine consumption. In their publications, writers Beno Kotruljević[25] and Nikola Vitovi de Gozze[26,27] witness the common abuse of wine beverages during the 15th and 16th centuries. Wine certainly represented the most important trade item in both villages and cities. The combination of countless sea and continental routes through which wine found its way in and out of medieval Dubrovnik ensured its well-being and prosperity.[28]

Special ordinances contained within the city statutes were designed to regulate the viticultural and winemaking activities. According to the Split statute, the vineyard had to be hoed twice a year and pruned once a year. If the tenant did not abide by the regulations, he had to give up that year's yield and pay a fine of four liberas (four pounds of silver). The proper cultivation of vineyards was ensured by severe daily fines of 20 soldas (3 grams of silver). In the case of neglect, the vineyard owner was whipped and banned from the city. The Dubrovnik statute clearly defined that the tenant's obligation was to thoroughly clean the vineyard according to the following instructions: "…for the first time by mid-March, for the second time by Vidovdan (June 28th) and afterwards as circumstances required.[29] Similar regulations about obligatory hoeing were contained within the statutes of Split, Hvar, Brač and Šibenik. The harvest was scheduled for mid-August.

In terms of wine types, medieval sources often only refer to white (*biancum*) or red (*vermelum*) wine. However, this does not mean that the classification of wine varieties did not exist. Documents accompanying wine imports to Dubrovnik clearly pointed to the grape variety from which the wine was made (*Maluasia*, *Romania*, *Maroa*, *Vernacia* and *Ribolli*). Malvasia Dubrovačka was a highly acclaimed variety, its wine often served as a delicacy, only in the high circles of the Republic.

The average price of wine in the 14th century can only be approximated based on a comparison of the different measures in use. At the time, about 80 liters of wine could have been purchased with one perper. The same amount of money would buy 24 kg of lamb or 47 kg of rye.[30] Thus, half a liter of wine approximately cost as much as 1 kg of rye. Considering the fact that rye was very expensive at the time, wine must have had a considerable price as well.

The rises and falls of the 19th century

It is safe to say that all the way until the 19th century, viticulture represented one of the most significant agricultural activities in Dalmatia and on the islands. The first ampelographic description of a portion of native varieties grown across the area of today's Republic of Croatia dates back from this period.[31] It is not precisely known how the author collected the data or who his helpers were, but Trummer's work is considered to be the oldest ampelographic source for some of the varieties that are still grown in Croatia today. Among others, it includes the descriptions for Plavanz, mali zrni (Plavac mali), Zherna belina (Mala modrina), and Heunisch ws. (Belina), varieties of considerable importance in the 19th century. In continuation of his work, Trummer visited vineyards of continental Croatia, and added descriptions for an additional 69 varieties from this region, accompanied by their corresponding German names. In the late 1850s, building upon the work of Trummer, Ljudevit and Novak continued with ampelographic descriptions of varieties from the south, including the renowned cultivars from the island of Hvar, such as Blatka bijela, Bogdanuša bijela, Crljenak crni (Blank blauer), Grk bijeli, Krivača bijela, Plavac mali and Dernekuša crna, etc. Stjepan Bulić compiled the single most complete ampelographic work, consisting of descriptions of 172 varieties (including the 28 Hvar described by Novak in 1887) indigenous to Dalmatia in his "Dalmatian ampelography".[10] This publication includes ampelographic analyses of native varieties grown in Dalmatia, many of which can still be found

Massive export of wine from Vela Luka on the island of Korčula, in 1901
(Source: Želimir Bašić - Dalmatinska vina kroz stoljeća, Šibenik 2001).

along the coast. The Ampelographic Atlas (I. and II.) published in 1952 and 1963, the work of Turković Z. and G. is a unique, detailed graphical representation of 60 table and wine grape varieties grown in Croatia. The most recent 2003 publication "Ampelographic Atlas", by Professor Nikola Mirošević and the late Zdenko Turković, is the most comprehensive compilation of the available ampelographic profiles for 150 (60 old and 90 new descriptions) grape cultivars grown in Croatia today. The descriptions are accompanied by graphic portrayals of the flower, petiolar sinus and cluster – the work of academic painter Ivana Gagić and the late Greta Turković.

In the first half of the 19th century, more attention was paid not only to biological and morphological characterization of grape varieties, but also to the improvement of viticultural practice, wine production techniques and quality. The production steadily increased and reached its "golden period" in the second half of the century, with improvements seen mostly in quantity and storage of wine products (in the medieval period wine was often spoiled by bacterial cultures). The first, or "small conjuncture" that lasted from 1850-1857, marked a sudden rise in wine prices and wine export to Lombardy and Venice.[32] This period was followed by an almost 30-year-long prosperity (1867-1892), or "big conjuncture", founded on territorial expansion of the vine and a large increase in production volume, resulting in bigger exports and greater profit. This is when Dalmatian wine became known to the European and world markets. According to certain sources[33,34], wine exports of Dalmatia were ranked between the fifth and seventh place in the world, behind France, Italy, Spain, Hungary and Portugal. Even so, the town of Split and its surroundings were compared to Oporto and Bordeaux, in terms of their importance in the wine trade.

Although at this point in time, thanks to the revenues collected from the wine trade, Dalmatia could have functioned as a self-sufficient state, the reality was not such. The Ausgleich (compromise) of February 1867 inaugurated a dualist Austro-Hungarian rule in place of the former unitary Austrian Empire (1804-1867). The new Austro-Hungarian Empire consisted of the Cisleithanian half and the Transleithanian half and included much of what is today Austria, Hungary, Croatia, Slovenia, and Bosnia-Herzegovina. As one of the "Kingdoms and Lands" of the Cisleithanian half of the Empire, Dalmatia was subject to the laws and regulations of the Austro-Hungarian politics headed by Emperor Franz Josef, who was also the King of Hungary.

It is safe to say that the vine was indeed the single most widespread crop in Dalmatia from 1850-1904; it occupied between one third and one fourth of the total agricultural areas of the region.[34] Dalmatia, according to Anton Dal Piaz, a renowned enologist from Vienna, was definitely the most important winegrowing region in the Austro-Hungarian Empire. The ratio of its viticultural and total geographic area in 1902, according to Ivo Juras, placed Dalmatia high among other viticultural regions in the world, second only to Italy. During the second half of the 19th century, the vine was not only the dominant crop, but also most evenly distributed and "knitted" into the lives of locals. "Dalmatia was created for the vine, and the vine was created for Dalmatia", recorded a source[35] from 1893 emphasizing the mutual dependence of the vine and the local population. At one point, more than 80% of Dalmatia depended on the vine and its products for survival.[36] In his marked speech at the Emperor's palace in 1908, Juraj Biankini, a fervent promoter of viticulture, warned about the potential adverse effects of such a symbiosis: "All the neighboring viticultural states, along with the vine, cultivate and make use of other agricultural crops. However, Dalmatia is completely focused on viticulture, and the survival of about 80% of its population depends exclusively on the vine".[37]

Although the two prosperous periods in Dalmatian viticulture seemed to have overcome the troubles from the past, they were predestined for an abrupt ending for several reasons. The farmers of Dalmatia were soon to be awakened from their dream and punished for relying only on the vine for daily survival. The boom in wine production and exports was based on lower quality, semi-finished products – bulk red wines intended for mixing with other wines (those of France, Italy, even Spain). Instead of being competitive in the field of high quality products and working on their diversity, Dalmatia had to build its fortune on the misfortunes of other wine producing countries. In the early 1850s, oidium had devastated Italian vineyards, followed by another pestilence, the phylloxera, in the 1860s. While the vineyards of France and other European countries were disappearing in the "claws" of phylloxera, Dalmatian winemaking prospered. Wine exports reached their highest levels, as phylloxera progressed on its trip across the European vineyards. Finally, the vineyards in the east of the Austro-Hungarian Empire succumbed, and the Hungarian market further opened up to Dalmatian wine in 1880. Dalmatia was unaware at this time that the same evil destiny awaited it only ten years down the road (significant sightings of phylloxera in Dalmatian vineyards were recorded in 1894). The viticultural collapse that followed had agricultural and demographic consequences from which Dalmatia has not recovered to this day.

The maximum in wine production was achieved in 1888; however, by 1889, the fire blight (*peronospera*) had already launched its attack on vineyards in Dalmatia and posed a serious threat to the farmers.[34] Around the same time, the Wine Clause from the Italian-Austro-Hungarian ship trade treaty of 1892, assured very low protectionist taxes for Italian wine. Dalmatian economy was betrayed by its own state as cheap Italian wine overflowed the markets. Just when the locals thought it could not get worse, after three decades of rampaging and plundering in Europe, phylloxera reached Dalmatian vineyards. For a long time these vineyards remained spared, mostly because its own poor infrastructure assured very scarce contacts between the numerous small vineyards scattered across Dalmatia. Also, the geographic isolation of the islands and individual growers on the islands was an advantage. For a while, it seemed that this pest would circumvent Dalmatia; however, the long years of resistance only guaranteed that the waiting and "expecting" had come to an end.

By 1898, approximately 25,000 acres (10,000 hectars) of vineyards were infected.[34] In 1899, phylloxera spread from Zadar to Benkovac and Šibenik counties. Approximately 17% of all viticulture areas (35,000 acres ≈ 14,000 ha) were affected by the end of 1900. The hardest hit was Split and its surroundings, where

Ruined vineyards of Dalmatia - a lost battle with fire during the 2003 summer months had devastating consequences on the island of Hvar. After decades of troubled viticulture, careless tourists and locals, aided by dry-spells, contributed to this sad sight (photo by Boris Kragić).

wine exports were the greatest. In 1888, the areas under vineyards in Croatia amounted to 430,000 acres (172,000 ha). By 1914, this number was reduced to 337,500 acres (135,000 ha) (250,000 acres in the coastal region and on the islands).[32] Partly due to neglect, but mostly due to lack of resources, by the time the bells rang for alarm, it was too late for the people of Dalmatia to rescue many of the autochthonous varieties and preserve them in a collection. As a consequence, the number of native varieties got reduced to one third (from approximately 400 to slightly over 100), and high quality cultivars such as Vugava and Malvasia dubrovačka were threatened by extinction. Without hope for recuperating from such devastation, almost overnight, people were forced to leave their homes in search of a better living.

The impact that the viticultural fall left on the demographic picture of Dalmatia was substantial, resulting in massive immigration of Dalmatian natives and ethnic Italians to overseas countries, especially America. According to one source,[37] at the turn of the 20th century 31,814 Dalmatians left their homes as a direct consequence of the final collapse of the backbone of the Dalmatian economy. In their speech held in 1902 at the Dalmatian parliament meeting in Zadar, alarmed with the situation that was getting out of hand, Kovačević and Biankini[38] declared that the 1892 Wine Clause is a crime against Dalmatia and its agrarian policy. As a consequence, Biankini added, "Dalmatian farmers are so desperate that most of them are willing to give up their estates and homes in exchange for an overseas trip to the Promised Land." At the Parliament meeting held in Zadar one year later,

Biankini again screamed: "This is a different kind of exodus happening in Dalmatia - entire families are departing for good, leaving nothing behind."[39]

Phylloxera was for the first time noticed on the island of Korčula quite late, in 1911.[40] It first appeared in a vineyard located in the Kruševo winegrowing region and from there spread across the other vineyards on the island. The production of wine on the island suddenly dropped, from 782 wagons (1876) to 100 wagons (1925). This stroke of misery caught the islanders by surprise. Hopeless of ever recuperating from such damage, they packed their belongings and headed overseas. It remained recorded that more than 3,000 inhabitants from the village of Blato left Korčula and most of them never returned: [41]

"Dear God, what a tragedy, how much pain, tears and sorrow. May it never repeat again!"

On another occasion, in 1925, about 1,500 people left the port of Prigradica in Blato, in only one day. According to the locals, the following song escorted them on their voyage:

> "Good bye my dear Prigradica cove
> Where my cheerful songs used to echo
> Good bye my dear village, my friend
> No time is left for us to spend " *

The depopulation of the village of Blato only mirrored the situation in many other Dalmatian towns and villages. Since the ending of the World War I and the disintegration of the Austro-Hungarian Empire in 1918, Dalmatia, as well as the remainder of today's Croatian territories, was a part of the Kingdom of Croats, Serbs and Slovenes, and later the Yugoslav state (from World War II until 1991 and the establishment of the independent Republic of Croatia). After half a century of battling vine diseases and pests, and experiencing historical upheavals, Dalmatia finally succumbed to the pressures. It was forced to export both its wine and its people and received almost nothing in return.

Throughout the passing centuries, the vine has been the essence of Dalmatia's being. Governing the lives of its inhabitants, it was their source of both prosperity and misery. During the day, the Dalmatian man left his sores, sweat and tears in the vineyards overlooking the sea; at nighttime, he rested and dreamed of the fall harvest. In prosperous times, the vine had kept families together working the land, only to force them apart, by a simple twist of fate, at another, less fortunate time. Without a doubt, the vine left a profound mark on Dalmatia's past - one that cannot be erased from the stone-carved memories of its people.

---

* Poem recited by Ivan Tvrdeić from Čara, free translation by J. Piljac

Dalmatian karst – a breeding ground for wine grapes

It is believed that the Mediterranean basin, due to its warm and relatively humid climate, has in the past played a particularly important role in the development of new wine grape varieties. The karst-crowned shores of the Mediterranean have been the breeding grounds, a sort of a world's nursery of cultivated grapevine varieties. Even today, the grapevine thrives in this region. The shores of the Adriatic in Croatia, with 1,185 islands, islets and reefs spread almost evenly along the coastal line, is where certain centuries old cultivars (Vugava bijela on the island Vis, Grk on the island Korčula, Malvasia dubrovačka near Dubrovnik), among more than 100 autochthonous varieties, are still grown today.

The wines produced in Dalmatia today are mostly varietals carrying either the variety name or the name of the growing region. Each region is, in turn, specified for only a few varieties that have, through the years, adapted to the growing conditions and managed to overcome nature's obstacles. The cultivars are chosen based on the following characteristics: resistance to diseases, an optimal ripening period, the total sugar and acid content of grapes as well as the ratio of other components (mineral, tannin, aromatic) that contribute to the overall harmony of taste, smell and color of wine. Thus, certain autochthonous varieties show positive qualities and result in the highest quality wines only at specific locations: Babić – Primošten, Vugava – Vis island (locality Podšpilje), Plavac mali – Pelješac peninsula and Hvar island (localities Dingač, Postup and Hvarske plaže), Grk – Korčula island (locality Lumbarda), Pošip – Korčula island (localities Čara and Smokvica), etc. The originality of wines produced from these regions is a consequence of not only the domesticated cultivars and the secrets of local winemakers, but the mixed influence of ecological factors such as climate, relief, the slope of the terrain, the proximity of the sea (diffraction of sunlight off the surface of the sea), the physico-chemical properties of the soil, as well as the richness of the surrounding vegetation.

The karst landscape of coastal Croatia, composed of rocks and sparse soil, has been a source of hardships and prosperity for Dalmatians for centuries. Living under hostile conditions and depending on small quantities of fertile soil for survival, the inhabitants of small villages and towns have learned to respect their land, be thankful for the yield it gives and be wary of its mutable nature. Learning to live on karst is a national necessity, as rocky topography prevails all along the 4,000 km long coastal line in Croatia. The term karst is a germanicized form of the Croatian words "krš" and/or "kras" which originally referred to the large area of the Croatian Dinaric karst and the distinctive limestone scenery in western Slovenia. The word "kras" appeared for the first time in 1230, on the island of Krk, in Northern Croatia.[21] The terminology that later referred to the karst in this region included "carso", "karst", and "causse". The most recently accepted term, still in use today in both scientific and popular circles, is "krš". It refers to the geomorphological, geological, pedological and hydrological characteristics of landscape and is used to describe the terrain with distinctive landforms and drainage caused by erosion.

There has been much debate over the conditions necessary for the development of karst terrain. Karst development requires an easily soluble rock, such as lime-

Vines thriving in the karst region of Kostanje - the hinterland of Split. Due to rocky morphology, soil had to be brought from other locations prior to the planting of vines (photo by Jasenka Piljac).

stone or dolomites. The weathering process of solution occurrs when natural waters come into contact with the rocks, leaving very little insoluble residue. An essential component of karst terrain is the development of an extensive underground drainage system, which leads to the production of a cave complex, as well as a range of closed depressions on the surface (e.g. valleys/dolina and fields/polje) through which water is chanelled underground. Erosion and deposition processes, superimposed on the karst terrain and maximized in humid climates, might be responsible for the distribution of *terra rossa* aggregations in various karst depressions. According to expert geologists, the accumulated red soil material has served as a base for the development of new soils suitable for the maintenance and continuation of centuries-old viticultural practices in Croatia.

The development of the karst terrain is also facilitated by the presence of considerably thick, well-jointed rock formations. In some parts of Croatia, carbonate rock units are over 3,000 m thick. Jurassic and Cretaceous periods were crucial for the development of such massive rock units. Since the end of the Eocene (35 millions of years-Ma), the External Dinarides (e.g. coastal Croatia) have been affected by karst processes and weathering, which have led to the development of distinctive karst landscapes, including the formation of different types of paleosols.

Geological composition of karst

An especially important layer in the composition of karst consists of sedimented dolomites, marl and limestone. Most of the development of lime-dolomite deposits took place Iin Creataceous and Jurassic periods. Karst is classified as deep and shallow. Deep karst is found only on the substrates of pure limestone, while shallow karst is layered upon dolomite rock foundations. In terms of viticulture, karst has certain advantages. The majority of the rainfall usually does not evaporate from the surface; rather, the water becomes stored in the underground layers from where deep-rooted vines easily reach the infiltrated moisture.

The main geomorphologic feature of coastal Croatia is an impressive Jurassic-Cretaceous carbonate succession, which palaeogeographically represents an isolated, intraoceanic Tethyan carbonate platform depositional area (or system), similar to the present-day Bahama bank. This succession consists predominately of limestone and sporadic dolomite layers.

Jurassic-Cretaceous carbonates of the coastal Dinarides (Croatian part of the karst Dinarides) were deposited in shallow-water platform environments. During the Jurassic period, the inner platform (present-day coastal Croatia) was characterized by stable and uniform sedimentary conditions (i.e. shallow marine environment separated from the land). This area was subjected to regional tectonism and sea level fluctuations, but the physiography of the shallow submarine area (platform) persisted throughout the Jurassic-Cretaceous and Early Paleogene (duration approx. 150 Ma).

Flysch deposits of coastal Croatia crop out in many small elongated basins: Pazin, Labin, Plomin, Ravni kotari, Central Dalmatia (Kaštela-Split) and Southern Dalmatia (Konavli).

The Kaštela-Split flysch deposits are generally characterized by an alternation of hemipelagic marls (predominated in succession) and hybrid carbonate-siliciclastic sandstones. The age of Dalmatian flysch corresponds mainly to Middle-Late Eocene (41-32 Ma).

The monotonous succession of marls and hybrid sandstones is randomly intercalated with several thick beds of breccias and/or conglomerates, i.e. megabeds (they show significant thickness of 0.5 - 5 m, sometimes over 10 m).

Several types of soil are found in scarce amounts in the Dalmatian karst region, all of which are suitable for viticulture. Sandy red soil (*terra rossa*), as well as all variations of brown, calcium carbonate-rich soils are considered especially favorable for winegrowing. Antropomorphic terraced terrains are common on the islands in the coastal belt and are classified as vitisoles on marly sandstone (Eocene flysch).[21] The deposits of all the aforementioned large-scale rock units, including flysch, are for the most part karst depressions (fields/polje, valleys/dolina) covered by quaternary *terra rossa*.

Terra rossa

The term *terra rossa*, commonly used in Dalmatia, refers to the red clay soil developed on carbonaceous rocks (mostly limestone and dolomites) associated with karst-specific features. Although it is commonly found throughout the Mediterranean, in Croatia, this soil is most represented on Dalmatian islands and along the coast with strong contrasts in precipitation. Dissolution of carbonates occurs during the wet season coupled with the release of clay and iron compounds as residual products. The strong red color is a consequence of conversion of hydrated ferric oxides to hematite during the dry season. *Terra rossa* is characterized by slightly alkaline to neutral pH and is almost completely saturated by calcium and/or magnesium.

The compositional features of *terra rossa* implied a close relationship between this type of soil and the underlying carbonates and, inherently, gave rise to a widely accepted theory that *terra rossa* has developed from the insoluble residue of carbonate rocks. The karst landscape of coastal Croatia is a typical example of the association between *terra rossa* and the underlying carbonates. In addition to the soil types common in the karst region, the climatic conditions of karst are also very important factors affecting the development of the vine and governing the processes of photosynthesis, respiration and transpiration in vine leaves.

Climatic characteristics of karst

Croatia has five climatic regions, two coastal and three continental.[42] Region B (shown on the map) includes the hills of Plešivica, Zagorje, Međimurje, Prigorje, Bilogora, Moslavina and Pokuplje with the average heat summation temperature (ΣEt) below 1370 °C. The most severe winters are experienced in northern Croatia from November until mid-March, with temperatures dropping below 0 °C during the night. The most abundant rainfalls take place during October and April-

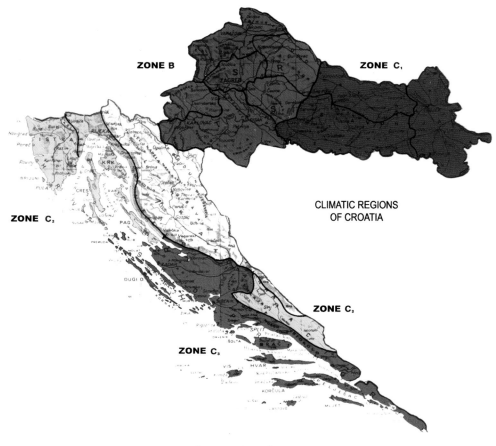

Climatic regions of Croatia
(Source: Fazinić Nikola and Fazinić Melita - Klimatske zone vinove loze u SR Hrvatskoj, Zagreb 1983).

May. Region C_1 consists of the plains and flatlands of Slavonija and Podunavlje with $\Sigma Et = 1371 - 1650$ °C. This is probably the most humid part of Croatia, where rain and rich soil contribute to ideal conditions for agricultural activities and the raising of both fruit and vegetable plantations. Region C_2 includes the Istrian peninsula, the Croatian northern coastal line and Dalmatinska Zagora, with $\Sigma Et = 1651 - 2204$ °C, where the Istrian peninsula boosts the average. The warm climatic conditions with mild winters, warm summers and karst-like landscape ensure favorable conditions for viticulture in this region. The temperature in the winter rarely drops below 0°C, and even if it does, it usually springs back up the next day. Region C_3 includes all of Dalmatia — northern, central and southern — all the way to Dubrovnik, with $\Sigma Et > 2000$ °C.

From the early beginnings, the vine had recognized Dalmatian islands, their karst-like landscape and extensive exposure to sunlight as its optimal habitat conditions. The islands account for 18.3% of the total geographic area of Croatia;[44] Their density is so high that the sea and the islands together comprise a unique aquatorium. Because of the variation in their geographical position, substantial climatic differences are observed between the coast and the islands, as well as among individual islands. Each island is a unique productive unit with different microclimate conditions, grapevine assortment, traditions and habits.

In general, the climate on the islands and the coastal belt is characterized by warm and dry summers, and mild and rainy winters. The average yearly temperature in the coastal region ranges from 16-17 °C with the lowest average temperatures reached in January (4-8 °C) and the highest in July (25-30 °C). The average yearly insolation on the islands and the coast ranges between 2,200 and 2,700 hours.[43] The average daily insolation recorded for the town Vela Luka on Korčula is 5.4 hours, while that recorded on Hvar is 7.4 hours. Such large differences in climatic conditions on the islands are quite common, and the resulting variation in ecological factors has aided the diversity of winegrape varieties grown in the karst region. The average rainfall falls in the range between 10.4 and 60 inches (260 and 1500 mm) per year. Just for illustration, the average rainfall on the island of Palagruža is 10.4 inches (260 mm) per year. On the island of Hvar this number is approximately more than two times greater, about 28 inches (700 mm) per year. As a consequence, vegetation varies drastically within a perimeter of only several kilometers, from abundant to sparse macchia consisting of evergreen rosemary bushes, cypress trees mixed with narcotic scents of garden sage, basil, laurel, colorful immortelles and lavender. All ingredients for a small herb pharmacy can be found within several square feet of rocky land. The variation in vegetation on the islands of Hvar (the island with the largest number of days exceeding 25 °C in the year) and Brač is so baffling, especially in early spring, indicating that their proximity gives no guarantees to the farmer and ensures only a quick trip from one island coast to the other.

It is evident that Dalmatian karst played an important role in the past, as very possibly the original home of numerous grape varieties, which later, via various trade routes, settled in other regions of Europe and the Americas. Although poor connections with overseas countries made the distribution of wine grapes somewhat harder prior to the development of air transportation, the sea route represented a slow but sure means of reaching the west. After phylloxera devastated Dalmatian vineyards, immigrants flocked to other grape-growing countries in hope of continuing their work under the more favorable conditions of California, Argentina, Chile, etc. Along with them they brought their old customs, crafts, knowledge and the culture of the grapevine.

All classic wine grapes have their roots somewhere in Europe and western Asia and it is a well-known fact that *Vitis Vinifera* L. is not indigenous to the Americas. With the aid of modern day science and technology, especially genetics, numerous well-known and economically important cultivars of today have been traced back to their origins in the Mediterranean and its surroundings. With the spreading application of the DNA fingerprinting technique combined with historical/ampelographic investigations, Professor Carole Meredith of the University of Cali-

Springtime on the island of Brač. A vineyard overlooking the village of Bol - a well-known tourist destination (photo by Boris Kragić).

fornia at Davis has investigated the origin of some of the most important cultivars that journeyed from Europe to the New World. Recently, the renowned grapevine geneticist and her scientific team have solved the parentage of Chardonnay[20] Cabernet Sauvignon[44], and Petite Sirah[45]. The exact origin of Zinfandel, one of the most important red wine cultivars of California, has been a mystery to the Americans ever since its arrival to the United States in the early 1820s. In 1998, Professor Meredith embarked on another challenge – solving the mystery of Zinfandel's origin.

CHAPTER TWO

THE STORY OF ZINFANDEL

Biological characteristics of the Zinfandel vine

Moderate vigour with upright shoots that mature to brittle canes characterizes Zinfandel.[1] The leaves are 5-lobed, medium to large, and fold inwards with small amounts of hair on the lower surface. The petiolar sinus is usually open and lyre shaped. Bunches are medium to large, more or less winged cylindrical and tightly packed with medium to large round berries. Zinfandel berries are easily distinguished from other varieties since most of them carry an irregularly shaped brown scar at the apex. Zinfandel has several viticultural problems, one of which is uneven ripening.[2] The bunch often ripens so that at the same time underripe, perfectly ripe, and overripe berries are found in a particular cluster. In addition, this variety is sensitive to bunch rot, spider mites and has a tendency to over-crop relative to vigor.[3,4] In order to avoid the chances of getting sunburned, it is advised to plant Zinfandel in the cooler districts. In comparison to other varieties, Zinfandel is considered to be a rather resistant and withstanding variety. Many Zinfandel vineyards more than 75 and 100 years old can still be found in California. It is believed that these old vines produce the highest quality wines because they give lower yields and their grapes tend to ripen more evenly.

Zinfandel is distinctive for the amount of secondary crop it is capable of yielding. At certain locations in Italy, a second harvest of Primitivo (Zinfandel) vineyards is quite common. When harvesting, special care should be taken because grapes are very sensitive to mechanical operations. In order to preserve unique varietal aroma, which is one of Zinfandel's greatest attributes, Zinfandel grapes are ordinarily not mixed with other varieties. However, in California, many wineries (like Ridge Vineyards) have become famous for their successful Zinfandel blends with Durif and Carignane.

The Zinfandel wine

Zinfandel makes a deep, full-bodied red wine, rich in fruit aromas and flavors. The impression of Zinfandel significantly changes and adopts a more serious tone with prolonged aging in oak barrels. Aromas extracted from the wood enrich the regular taste profile of the fruit. Zinfandel grapes are interesting because they can be made into almost every possible wine style, ranging from strong, dry and tannin-rich reds (up to 17% alcohol) to the easy drinkable ones (14% alcohol), on to rosés and "white" Zinfandels. In its light and fruity form, Zinfandel reminds one of the French Beaujolais; in its robust and aged form, it is most similar to Cabernet.

A close-up of Zinfandel, from which the characteristic uneven ripening of grapes is evident (photo by Jasenka Piljac)

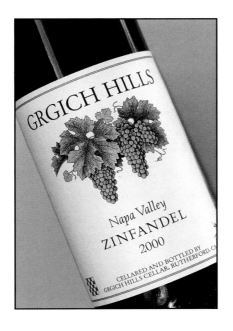
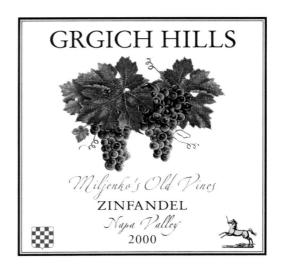

Grgich-Hills Zinfandel (Source: Mike Grgich).

Zinfandel grapes are also used to make late harvest wines with residual sugars and sweet dessert wines, high in alcohol (portlike). California Zinfandels typically have a strong raspberry, black cherry or blackberry taste, mixed with a variety of flavors extracted from oak barrels. These wines are often cellared for a short time (5-8 years) after bottling to achieve optimal properties. Zinfandel's peculiar and unique uneven ripening process, whereby the cluster consists of slightly underripe berries, perfectly ripe and dehydrated ones, adds to the flavor complexity of the wine. In the winemaking process, acidity is extracted from the underripe berries, berry flavors and fullness from the ripe and overripe berries.

Zinfandel wines are usually served with grilled meat and are compatible with most red meat dishes. White Zinfandels are produced by picking the grapes when the sugar levels are low (18-21 %) and minimizing the contact of skins with the juice during fermentation, so that only a portion of the pigments get extracted. These wines are suitable for lighter meals and are often drunk young, after only a few months of ageing when there is some residual sugar left. Changing style with Zinfandel always assumes adding variety but keeping the quality.

Zinfandel's popularity

Zinfandel has been widely grown in California since the 1800s. It was produced in the Lodi area as early as the 1860s. Owing to the encouragement of commissioner Charles Wetmore, the plantings of Zinfandel in Lodi spread even more in the 1880s. White Zinfandel, the lighter version of the original red, was marketed for the first time in the 1960s by David Bruce. Following this, Sutter Home Winery popularized white Zinfandel and launched a six million case a year market in the 1980s. Today, more than 300 producers grow Zinfandel over 50,000 acres in California at locations including Lodi, Contra Costa, Mendocino, Paso Robles, Napa, San Joaquin, San Luis Obispo, and Sonoma counties.[5] A lot of Zinfandel fruit in California is used for white Zinfandel, but red Zinfandel is again gaining in popularity and is the only type of Zinfandel consumed by wine *connoisseurs*.

The importance of Zinfandel in the US wine industry is evident, as it is considered to be one of America's most important red wine grapes. According to the California Agricultural Statistics Service[6], in 2000, Zinfandel was planted on slightly more than 47,000 acres ranking second place among the red wine cultivars, behind Cabernet Sauvignon (48,285 acres). In the Final Grape Crush Report for 2001, Zinfandel shared second place with French Columbard behind Cabernet Sauvignon in terms of the percent of total crushed crop (10%). In the report for 2002, Zinfandel and Cabernet Sauvignon again emerged as the leading red wine grape varieties (9.8 and 10% of the total crushed crop, respectively), leaving French Columbard behind (8.3%).

California is the leading wine producing state in America; it accounts for approximately 90% of all US wine production. If California were a nation, it would be the fourth leading country in the winemaking world, behind France, Italy and Spain. In such a developed wine industry of California, Zinfandel accounts for roughly twelve percent of the wine sold in the US each year. Translated to dollars, Zinfandel's share of the market for domestically produced and consumed wines significantly exceeds $10 million.[7]

In California, Zinfandel enjoys almost a cult-like following. The largest non-profit organization in the wine world, The Association of Zinfandel Advocates and Producers (ZAP), is dedicated to the promotion of Zinfandel and education of the public about this economically important variety. Zinfandel Advocates and Producers, founded in 1991, consists of about 6500 advocates and over 310 producers of Zinfandel. According to ZAP, Zinfandel is cultivated in Arizona, California, Colorado, Illinois, Indiana, Iowa, Massachusetts, Nevada, New Mexico, North Carolina, Ohio, Oregon, Tennessee and Texas. It can also be found in Australia, South Africa, New Zealand, France, Italy, even Chile. The organization directly supports the research related to Zinfandel conducted at the University of California at Davis, including: the development of the Zinfandel wine aroma wheel; the collection of century-old vines across California and their preservation in the Heritage vineyard located in Oakville in the Napa Valley; and the DNA analysis of Zinfandel vines from the Heritage vineyard. ZAP represents a link between the scientists, expert winemakers and the public with its annual activities that bring together all three groups. With a twelve-year tradition, the Annual Zinfandel Festival is held each year in January. This five-day long event is a unique opportunity for the wineries to pour their Zinfandels and the public to see what's new on the market. In the last Zinfandel festival, approximately 10,000 visitors participated in the festivities. This event is most probably the world's largest tasting of only one varietal. ZAP is also the organizer of Zinfandel Days in each of California's

Grgich-Hills Zinfandel vineyards in Calistoga (Source: Mike Grgich).

winegrowing regions, specialized tastings, seminars and educational programs across the US.

Mysterious origins - the American view

Professor Charles L. Sullivan, a historian who has dedicated years of his work to researching California's viticultural practices, once referred to Zinfandel as "…the spirit of American pioneers captured in the bottle for all times."[8] Many American wine-lovers identify with the hardships of this wine, on its road to success and popularity. The fact is that this variety was practically nonexistent, let alone famous, in the winemaking industry until its arrival to the US. Although it is widely accepted that all *vinifera* varieties originate either in Europe or western Asia, many argue that the name Zinfandel, the taste of the wine, and the eco-geographical zone of production in California are unique and cannot be found anywhere else in the world. Thus, the Zinfandel wine is no less American than the soil from which it draws its essential moisture and nutrients.

In the light of numerous investigations aimed at uncovering Zinfandel's European origins, California winemakers felt threatened and alarmed. Their concern spread and even reached the state authorities in 1999. In an effort to protect this California treasure and encourage its awareness, in July of 1999, California senators Barbara Boxer and Diane Feinstein introduced Senate resolution 132[9] designating the week beginning January 21, 2001, as "Zinfandel grape appreciation week". This resolution listed the merits of the Zinfandel grape and its significance among agricultural products of the United States, concluding with the statement that the "Zinfandel grape is a national treasure". Upon evaluation by the Judiciary committee, the bill did not come up for a vote. The same year, the Bureau of Alcohol Tobacco and Firearms (BATF), the ruling body on alcohol classification, proclaimed Zinfandel the only wine grape varietal considered to be unique to the US.

Without a doubt, Americans have recognized the potential of Zinfandel, turned it into a high quality product, and brought it to the attention of the rest of the world. Zinfandel's road to prosperity was hard. It had to journey from Europe to the American east coast, from the east coast on to California where it finally established itself and earned the title "one of America's most important wine grape varieties". However, since none of the *Vitis vinifera* L. varieties are native to the Americas, Zinfandel's European origins continued to be as mysterious and intriguing as the red liquid that enchanted the world.

A chronology of the European origins of Zinfandel

A nobleman's grape?

For a very long time, the only link between Zinfandel and its European roots was believed to be Agoston Haraszthy, a Hungarian immigrant who planted over 300 different varieties and over 100,000 cuttings of grapes in California. Haraszthy was born in 1812 in Bacska County, Hungary, to the nobleman Charles Haraszthy de Bacska and Anna nee Halasz.[10] He was appointed to the Royal Hungarian Guards of Francis I, Emperor of Austria-Hungary in 1830. Because of his close relationships with the Transylvania reformer, Baron Wesselenyi and the Hungarian revolutionary hero, Louis Kossuth, Haraszthy was politically confronted by the Austrian Emperor. After witnessing his political allies being charged with treason in 1837, and fearing for his life, Haraszthy decided to seek refuge in America. He traveled through Ohio, Indiana, Illinois, Wisconsin, Iowa and Kansas. The vast landscapes and untouched, fertile land impressed Haraszthy and reminded him of his native Hungary. He recognized the hidden potential of newly discovered territory and purchased 10,000 acres from the US Government along the Wisconsin river and later moved his entire family to Wisconsin in 1842.

Haraszthy's dream was to plant high quality vineyards in America and introduce to his new homeland the viticultural tradition of Hungary. He planted the first large vineyard at Crystal Springs in San Mateo County, California. In 1857, Haraszthy visited Sonoma Valley and found its climate, topography and soil ideal for winegrowing. In Sonoma, Buena Vista, he established an estate, started his own winery and made wine until the late 1850's.

Among his many political duties, Haraszthy even served as the San Diego county sheriff from 1850-1851. In 1861, he was appointed to the California commission with a task to improve agriculture methods, and enrich the California fruit collection with new vines and fruit tree stocks from Europe. During his five month European tour with son Arpad, Haraszthy purchased 100,000 grapevines representing 1,400 established European varieties, along with selected stocks for olives, almonds, oranges, chestnuts, lemons and pomegranates.[10] Most vines were planted in Buena Vista, and financially supported by the Buena Vista Viticultural Society.

Partly because of his political and agricultural endeavors and partly because of his interesting personal life path, several publications have been dedicated to Haraszthy.[10,11] His adventurous life ended in Nicaragua where he moved after the Buena Vista Horticultural Society accumulated debts. Agoston Haraszthy died on July 6, 1869, near his estate Hacienda San Antonio at Corinto Nicaragua, while attempting to cross a river full of alligators. His body was never found.

Agoston Haraszthy, pictured in New York City in 1861, right before his trip to Europe (Source: Charles L. Sullivan).

For a very long time, Agoston's son Arpad promoted the idea that Zinfandel came from Haraszthy's collection; however, investigations of Professor Charles Sullivan proved otherwise. In his recent publication[12] that compiles the findings of research conducted over the course of the past several decades and is aimed at reconstructing Zinfandel's history, Professor Sullivan dedicated an entire chapter to debunking the Haraszthy myth. Sullivan's extensive historical investigations, supported by bountiful evidence, confirm that the credit for Zinfandel's arrival in the US was wrongfully given to Agoston Haraszthy and later on further perpetuated by his son Arpad.

According to Sullivan,[8] the grape arrived as a nameless vine from the Schönbrunn imperial collection in Vienna, which contained all vine varieties grown in the regions of the Austrian Empire. A Long Island grape grower, George Gibbs, imported it somewhere between 1822 and 1829 for his nursery. Soon after, along with the other grapes, Zinfandel was appearing at local agricultural and garden fairs, but under a different name. In a catalogue of another nursery in Long Island, it first appeared in 1829 under the name "Zinfardel". In 1832, a Boston nurseryman, Samuel Perkins, advertised the "Zinfendal" vine for sale,[5] and in the years that followed, the grapes were sold under the names "Zinfindal" and "Zinfendel" throughout the northeast and the east coast, from Long Island to Boston, where most of the *Vinifera* grapes were grown under glass. In 1846, J.F. Allen,[5] a New England expert on viticulture, described Zinfandel in America's first academic book on viticulture. It is well known that a variety morphologically identical to Zinfandel under the name Black St. Peter's was grown in New England greenhouses

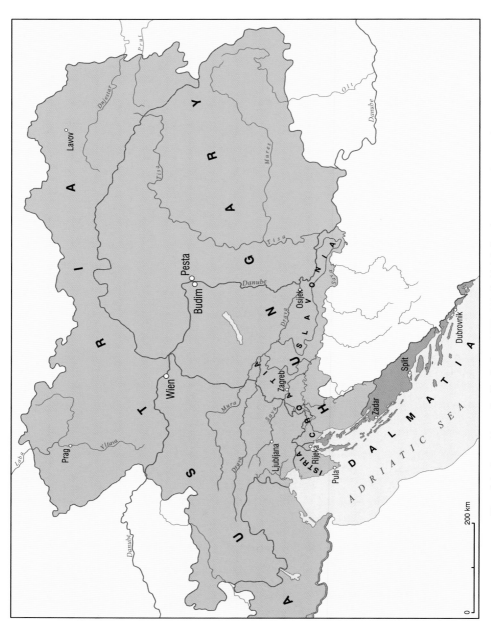

Map of the Austro-Hungarian Empire from 1867-1918 of which Dalmatia was a member-province (drawn by Tomislav Kaniški).

at that time, but the connection between the two variety names could not be established.[13] According to the data from one nursery catalogue, Zinfandel was imported from Germany. However, in the middle of the 19th century, a country by the name of Germany did not exist – this detail could only mean that the vine was imported from a German-speaking region, from Vienna in the Austro-Hungarian Empire.

The trip to America

Zinfandel arrived via the sea route to California, most probably in 1852, right after the beginning of the Gold Rush. Captain Frederick Macondray, a fervent grape grower, brought to his friend in Napa several vine cuttings, among them, Traminer, Muscat d'alexandrie and the lesser known Zinfandel.[12] Several years later, the captain's friend sold two wagons of Zinfandel cuttings to the Horticultural society in Sonoma. Most of the vines froze that year, but Zinfandel survived. When the first Zinfandel wines were made in California, they were taken to the General Vallejo winery in Sonoma. The French winemaker, Victor Fauré, tasted the wine and to everyone's surprise, declared it excellent, very similar to the French *claret* (*claret* refers to the red wines made in Bordeaux style). In the early 1860s, the grape name was finally agreed to be Zinfandel, and the variety was proclaimed an excellent one for the production of red table wine in the Sonoma and Santa Clara counties, and the Sierra Foothills. However, its popularity throughout northern California spread much later, in the late 1870s and 1880s. Between 1878 and 1889, Zinfandel was the most widely planted variety in California, occupying thousands of acres. This period marked the beginning of the California wine industry; in the late 1870s the most reputable winegrowing region in California today, Napa Valley, had 3,000 acres under cultivation as vineyards. At the end of the viticultural boom, Napa Valley spread its vineyards across 20,000 acres.

Between 1890 and 1900, most of northern California's vineyards were destroyed by phylloxera. Zinfandel was among the lucky varieties planted on resistant rootstocks that survived this viticultural plague. In the period that followed, California vineyards were replanted, and Zinfandel emerged as a leading varietal in the production of red wine blends. Zinfandel wine was often blended with Cabernet, Petite sirah, Carignane or Mourverde; only a few wineries produced the varietal Zinfandel. Large companies in San Francisco bought it from smaller producers and sold it under the name "claret". Before the Prohibition, the Gundlach-Bundschu winery in Sonoma and Krug and Jacob Schram stood out from among the other Zinfandel producers.

During the Prohibition (1920s), it was illegal to sell wine in America; however, Zinfandel remained popular among the small home winemakers and was one of the top five varieties in the country. The popularity of Zinfandel continued to grow throughout the next three decades and high quality varietal Zinfandel wines became common and sought-after...

First modern traces in Italy

The connection between Zinfandel and the Puglia (Apulia) region of Italy was made in the late 1960's.[14] In the autumn of 1967, Dr. Austin Goheen, a University of California at Davis researcher, visited Bari located at the heel of southern Italy. He tried the local Primitivo wine, which reminded him of Zinfandel, and asked to be taken to the local vineyards. The resemblance between Primitivo di Gioia and Zinfandel vines prompted Goheen to take cuttings of Primitivo back to Davis where the vines were compared in side-by-side plantings with Zinfandel. Identical morphological characteristics were observed. At that time, sophisticated DNA analyses were not yet available and only isozyme (enzymes that are unique to each variety) patterns of the vines could be compared. Wade Wolfe, a doctoral candidate at UC Davis, performed the isozyme analysis in 1975 and proclaimed that, as far as the latest scientific techniques allowed him to conclude, Zinfandel and Primitivo are indeed identical. Although the results of isozyme analyses cannot always be taken with a 100% certainty, the wine-loving public of California soon initiated comparison tastings of Italian Primitivos and Californian Zinfandels. Comments followed that "the Primitivo could well be Zinfandel"[15] and soon afterwards, bottles of Italian Primitivo found their way in cases of wine labeled as Zinfandel. California Zinfandel producers and consumers felt betrayed and, helped by the media, initiated an outpouring of newspaper and magazine articles whose aim was to protect the American heritage of Zinfandel wines.

In 1985, the BATF ruled that "Zinfandel" may not be used as a synonym for "Primitivo", because (i) there was not enough evidence that proved that Zinfandel and Primitivo are "one and the same"; and (ii) Primitivo wines could not be sold under the name "Zinfandel" in Italy since the EU's list of Italian grapes did not include Zinfandel. For a while, the California Zinfandel producers seemed satisfied; however, Professor Meredith was determined to find out the scientific truth about these two varieties. Successful application of DNA fingerprinting techniques on grapes, performed by her graduate student, John Bowers, had made possible the comparison of the DNA profiles of Primitivo and Zinfandel. The results were consistent with the previously performed isozyme analysis and it was concluded that Primitivo and Zinfandel are clearly the same variety.

This finding has again rekindled a controversy in wine circles, as the Italian wine producers now felt that they were entitled to sell their Primitivo wines under the label Zinfandel.[7] After all, wines made from the same grape should be called the same, the Italian Primitivo producers argued. Besides, in the case of Zinfandel, Italian producers had a legitimate argument since Americans had been using the common names of foreign varieties such as Cabernet Sauvignon and Merlot. American Zinfandel producers, however, pointed out that the taste profile of Apulian Primitivo is not even close to California Zinfandels, and that the selling of lower quality Primitivo wines as Zinfandels would have an adverse effect on Zinfandel's reputation. American producers were reluctant to "lend" the Zinfandel name to their Italian colleagues, firmly believing that this name belongs exclusively to California-produced grapes and wines nurtured with a different style. In Italy, Primitivo is often associated with bulk wine, because it easily achieves high alcohol contents, and it is an unknown on the international market. The time and money invested in building the reputation for Zinfandel in California and abroad were also an issue; besides, California winemakers were concerned that Italian Zinfandels might pose a threat to their market in the United States and the European Union.

In January of 1999, the European Union granted Italian Primitivo growers permission to use the name Zinfandel, and in July 2000, the California winemakers responded by filing a complaint against the Italian Zinfandel producers with the Bureau of Alcohol, Tobacco and Firearms.[7] They also turned to ZAP for help; however, there were no legal grounds on account of which the 1985 BATF ruling could be enforced. As far as the Italian Primitivo/Zinfandel producers were concerned, without a requirement for a special announcement, scientific facts had just burried the 1985 BATF ruling. The complaints and lobbying against Italian Zinfandel producers have not been successful thus far, and it seems that the Italian and Californian winemakers will have to find a common ground in the end.

The Puglia region is traditionally known for its powerful red wines with 14°-15° of alcohol and impressive rosés made from the Primitivo, Malvasia nera and Negroamaro grapes. The "old-style" Primitivo wines are characterized by red-brownish pigmentation and baked fruit flavors; however, ever since the discovery that Zinfandel and Primitivo are identical, the Primitivo producers are exploring the possibilities of changing the style of Primitivo towards the lighter, and internationally more acceptable purple pigmented, fruity wine. Gregorio Perrucci, an Apulian winemaker, has recently assembled Apulian wine producers into an association under the name Accademia dei Racemi, whose aim is to pursue high quality wine production with the main focus on Primitivo. In his effort to explore the potential of Primitivo, Gregorio Perrucci realized that he will have to look at the quality factors associated with the Puglian winegrowing regions and specific localities. However, since there is no written history of the Primitivo grape in Italy, most of the information related to the tradition of Primitivo can only be obtained from the old generations of grape farmers and winemakers.

The old vineyards of Primitivo in Puglia are located in proximity to the seashore, in red soils. Since phylloxera could not easily survive in this type of soil, composed of sand, iron oxide and calcium, the vines there are on their own roots. The composition of soil is similar to the one found in southern Dalmatia; the climate conditions of these two regions also overlap (fourth climatic region according to Winkler[16]). The closeness of the sea provides humidity and assures cooler conditions during hot summers. The moisture gets trapped between the sand and rocks, where the deep-rooted vines can easily reach it. In southern Dalmatia today, certain old vineyards, especially on the islands, are also own-rooted. It is clear that the viticultural regions of southern Dalmatia and Puglia could easily host the same grape varieties.

Although the finding that Zinfandel and Primitivo are identical has narrowed down the search for the original home of Zinfandel to the Mediterranean basin, its exact origins were still debatable. The documented evidence of Primitivo's presence in Italy extends back only to the mid-1700s, and although this does not imply that Primitivo's presence in Italy runs only 300 years, the hypothesis that it is not a native Italian grape seemed plausible. The geographic proximity of Croatia, its long viticultural tradition, as well as viticultural conditions resembling those of Puglia, pointed to this country as its possible original home.

Zinfandel in Croatia

The idea that Zinfandel originates from coastal Croatia, and that it is the same as Plavac mali (the most important red wine cultivar of this region), has been drifting in scientific circles for at least 20 years. In his publication "Wines of America",[17] Leon Adams wrote of a grape, indigenous to the Dalmatian coast of Croatia, whose description is very similar to that of Zinfandel. Professor Sullivan later learned that the first cuttings of Plavac mali arrived on the Davis campus as a consequence of extensive correspondence between Dr. Goheen and Professor Ana Šarić of the University of Zagreb, who worked with native Croatian grape varieties. The analysis of isozyme patterns performed by Wade Wolfe for Plavac mali and Zinfandel, however, led him to the conclusion that the two varieties exhibit certain differences.[18] Nevertheless, numerous viticultural experts in Croatia still condoned the hypothesis that Plavac mali is the Croatian counterpart of Zinfandel.

In his study of the origin of Plavac mali, Maleš[19] wrote that this cultivar "was carried over to the Italian coast of the Adriatic sea, Puglia-Bari (where it is called Primitivo) and to the west coast of the USA (where it is called Zinfandel)." Mike Grgich, a well-known Californian winemaker of Croatian origin, was the motivating force behind the idea that Plavac mali is the same as Zinfandel.

Plavac mali is the most important red cultivar of middle and south Dalmatia, best adapted to the south-facing locations of the islands and the coast, especially Dingač on the Pelješac peninsula. The first written records pointing to the presence of Plavac mali in Damatia come from Trummer.[20] In his Dalmatian ampelography, Bulić[21] listed a large number of synonyms used for this cultivar in various winegrowing districts of Dalmatia. In the available literature, there is no mention of Plavac mali being grown under this name anywhere outside of Croatia.

The genocentre of Plavac mali is relatively narrow; today, this variety is grown in the short coastal strip from Konavle to Primošten with the main centers on the Pelješac peninsula and the islands of Hvar, Vis, Korčula and Brač. A hilly-mountainous relief composed of cretaceous limestone and dolomites, eocene limestone and marl, characterizes the coastal area where Plavac mali is grown.[19] The genetically predetermined potential of Plavac mali grapes becomes most pronounced on the south-western positions of the Pelješac peninsula, the Postup and Dingač appellations. Here, the slopey vineyards descend towards the sea almost at a 90% angle, and in the ripening process, the grapes receive a unique mixture of the three essential elements: earth, air, and water. The mean monthly value of air temperature during the vegetation period (21.3 °C) and the mean annual value (16 °C)

Plavac mali - the major red wine grape cultivar of the Dalmatian coast of Croatia (photo by Boris Kragić).

are ideal for the ripening of grapes and maturing of wine on the Pelješac peninsula.

Plavac mali grapes from the Dingač and Postup appellations are made into top-quality wines. The Dingač wine was the very first superior quality red wine with protected geographical origin in Croatia. It is one of the well-known exports of Croatia, and the most famous Croatian wine available abroad. The Dingač wine is full-bodied, slightly dry and astringent, with a distinctive Plavac aroma contributing to its luxurious bouquet. In general, wines made from the Plavac mali grapes are characterized by a deep ruby-red color, a warm-rounded bouquet and variety-specific aroma. The processing technology of Plavac mali grapes is based on the quick extraction of color and prevention of excessive secretion of phenols.[22] Most Dingač wines are produced by small private wineries that either own vineyards in Dingač, or buy the top-quality Dingač grapes. The largest single producer of the highest quality Dingač wine is a cooperative winery in Potomje, Pelješac. While Plavac mali wines average between 11.5-14.5 % alcohol vol, the Dingač wine may reach up to 17.6% alcohol vol.

The controversy surrounding the Croatian origin of Zinfandel sparked an interest in the scientific circles of California. In an attempt to initiate the investigation of Plavac mali in Croatia, firmly believing that this variety deserves the attention of grape experts, Mike Grgich contacted the renowned "grapevine detective", Professor Carole Meredith. With numerous grape mysteries now solved and behind her,[23,24,25] she accepted Zinfandel as another challenge, probably the greatest one in her entire carrier. Coincidentally, at the same time, I was offered a

Plavac mali vineyards in Dingač - the south-facing sloped terrain on the Pelješac peninsula (photo by Boris Kragić).

student-assistant position in her laboratory and our paths crossed in Wickson Hall for the first time.

As Professor Meredith later concluded, pure serendipity brought me to her laboratory. I joined the Meredith team at UC Davis in 1997, as a senior year biochemistry student looking at earning some extra money prior to returning to my homeland, Croatia. I had very little practical laboratory knowledge and felt most comfortable washing the dishes and observing the graduate students do the high science. Little did I know at the time that of all the job offers posted at the Student employment center, I picked the one that will lead me back to Professor Meredith's laboratory many more times and that my accidental choice will grow into a fortuitous scientific and personal experience.

The year later, in the spring of 1998, I came back to Davis and the laboratory supervisor, Gerald Dangl, encouraged me to participate in the scientific work carried out in the lab. In a couple of months, I learned the technique of microsatellite analysis that consists of comparing specific repetitive stretches in the DNA of grape varieties in order to identify and determine the relatedness between them. In the beginning, I worked mostly on identifying certain unknown varieties from California vineyards. During my second stay in Davis, Professor Meredith informed me about her plans to visit Croatia and start collaboration with the scientists from the Faculty of Agronomy at the University of Zagreb. Namely, the Plavac mali material kept at Davis was not sufficient for any kind of a serious study and more original Plavac DNA had to be collected. The two available accessions at Davis came from two different sources, one from a collection in Italy and the other form a nursery in Zagreb.

The Dingač and Postup wines produced by one of the many wine cooperatives on the Pelješac peninsula, PZ Potomje (Source: Several authors - Hrvatska vina i vinari, Zagreb 2002).

I was told that Zinfandel's traces have long been pointing to Croatia, but it seemed that until then, no one had the means, or the necessary motivation, to begin the search in Croatia, a country that has been ravaged by war from 1991-1995. She offered me the opportunity to be her companion and translator. I felt honored by this suggestion, seeing it as a challenge that would let me learn science from the best grape geneticist in the world and, at the same time, as an opportunity to promote my homeland. I was looking forward to the excitement ahead of us and, in preparation for the trip, worked on translating the ampelographic descriptions of most interesting Croatian varieties and getting myself familiarized with the "Zinfandel story".

Just about the time when Professor Meredith decided to begin her search in Croatia,

The beginning of Zinquest - this picture of Professor Carole Meredith and Jasenka Piljac was taken in May of 1998, right after the meeting with the presidential counsel for economic affairs, Lidija Zorić, at the Offices of the President of Croatia in Zagreb (Source: Jasenka Piljac).

scientists, Dr. Ivan Pejić and Dr. Edi Maletić from the Faculty of Agronomy, informed her of their new project whose aim was genetic characterization of autochthonous Croatian grape varieties using novel molecular methods. As a country-candidate for entrance into the European Union, Croatia was given the guidelines according to which it had to meet up to the standards set by other countries in the European Union. Croatian scientists realized that little time (until 2007) was allocated for a thorough inventory necessary in the viticulture department (the exact determination of the number of native varieties, development of a wine land register, etc.). The motivation behind their quick action was the fact the European Union regulations restrict the further expansion of viticultural areas for member-states. The present state of viticulture in Croatia rang an alarm. The timing could not be better and the two sides easily found the common ground. Like pieces of a puzzle, all the components necessary for a successful trip to Croatia fell into place. Professor Meredith arranged a meeting in Zagreb, while Pejić and Maletić promised to help her on the "Zinquest" by contacting the local grape growers and finding guides that will help us tour the Plavac mali vineyards. Before leaving for Dalmatia, she scheduled an informative lecture for the scientists at the Faculty of Agronomy that marked the beginning of a new collaboration with one common aim - finding out the truth about Zinfandel.

The Plavac mali controversy

Our island-hopping tour in Dalmatia, in early May of 1998, lasted two weeks. The objective was to investigate as many different Plavac mali vineyards as possible, especially those that might contain old plant material. With the help of new collaborators from Zagreb, leaf samples were taken from 148 individual Plavac mali vines grown in 45 different vineyards on the Pelješac peninsula, the islands of Hvar and Korčula located just across from Split, a major southern Dalmatian port. The sampling was based on the initial hypothesis that Plavac mali and Zinfandel are one and the same variety; thus, the search was targeting those vines which morphologically resembled either of the two cultivars. During the trip, we made careful recordings of the locations and vineyards we visited so that later it might be possible to go back to a specific location, if necessary. Upon returning to Zagreb, I was able to reconstruct a portion of our trip according to my notes, as follows.

Arriving in Trstenik

The first stop on our Dalmatian tour was Mike Grgich's winery in Trstenik on the Pelješac peninsula. Grgich left native Croatia and came to Napa Valley of California in 1958 where he established a winery in 1977 in partnership with Austin Hills. Even before establishing his own winery, as an employee of Chateau Montelena, Grgich became respected in the Californian winemaking circles after producing the award-winning 1976 Chardonnay. As a great patriot and wine enthu-

Leaving Split - a panoramic view of the Split port where we boarded the ferry to Hvar in the first search of the territory in 1998 (photo by Jasenka Piljac).

Mike Grgich's winery in Trstenik on the Pelješac peninsula (photo by Jasenka Piljac)

siast, Grgich eagerly awaited the beginning of the "Zinquest". He offered Professor Meredith accommodations in Trstenik throughout the duration of her visit to Croatia. The years of Grgich's lobbying and persuasion in California had finally paid off with Meredith's initiative and trip to Dalmatia. Zinfandel's mystery was soon to be uncovered with the last words in the book on its origins written with only four letters of the genetic alphabet – A, T, G and C.

The Grgich winery in Trstenik was opened in the mid-1990s, when Mike Grgich purchased a large building previously owned by the government and redesigned its interior into a modern wine cellar. For a long time, Grgich's wish was to invest his knowledge and money back into Croatia and make the highest quality wine in the region, according to his California recipes. He chose Plavac mali from the Pelješac peninsula and Pošip bijeli from the island of Korčula for this task. Today, the winery in Trstenik is renowned for its top quality Plavac mali and Pošip wines. The Plavac mali wine is characterized by vanilla aromas extracted from barrique, blended with the earthy taste of Plavac grapes from the Dingač and Postup appellations. This high alcohol wine (up to 17 vol %) is probably one of the best reds on the Croatian market. Grgich's Pošip, renowned for its unique varietal aroma, is a wonderful example of a noble white wine suitable for complementing Dalmatian fish specialties.

We spent our first day on Pelješac sightseeing Trstenik and marveling at the beauty of Dingač and Postup vineyards, located on the steep terrains facing the sea. Grgich organized a boat ride with one of his local friends, Jozo Braskin, and, although the day was cloudy, perfectly aligned vines growing on the western slopes of the peninsula filled everyone with admiration. We learned that the grapes in Dingač achieve higher contents of sugar in comparison to Plavac mali grapes at other localities, because they absorb direct sunlight as well as the rays that reflect off of the sea surface. Due to the unique nature and quality of wine produced from grapes in this region, the Dingač wine was Croatia's first red wine with protected geographical origin.

On a boat ride across the sea from Trstenik, from where the steep vineyards of Postup and Dingač could best be viewed. From left to right: Mike Grgich, Ivana Popović, Jozo Braskin and Professor Carole Meredith (photo by Jasenka Piljac).

After completing the boat tour, Jozo invited everyone to his home for dinner. Trstenik is a small village numbering only a few hundred inhabitants throughout the year for whom, except for fishery and winegrowing, tourism represents the main activity in the summer months. That evening, Jozo decided to demonstrate his expertise as both a fisherman and a cook. We were joined at dinner by Jozo's wife, daughter and granddaughter Ivana who arrived the day before from Dubrovnik. Together with his wife, Jozo prepared fresh mussels in wine and tomato sauce, a Dalmatian specialty, with added spices of rosemary, basil and bay leaves. The dinner was accompanied by homemade bread and Jozo's Plavac, made from the grapes we previously saw from the boat. Jozo's stories about winegrowing and life in Dalmatia, and his pride in being able to tell them to Professor Meredith, added to the unique atmosphere at the table. Often, I could not translate in time everything Jozo was trying to communicate in his excitement. We were impressed with his openness and simplicity, and the ease with which his family carries on the struggle with the sea and the land.

A memorable evening marked the beginning of our search for Zinfandel.

The antique beauty of Hvar

We began the search on the island of Hvar. Upon arriving in Split, after a four-hour drive from Trstenik, we met with the scientists from Zagreb, Ivan Pejić and Edi

Maletić. The meeting was previously arranged in Zagreb at the Faculty of Agronomy, since their help was needed in locating and searching the local vineyards of native Plavac mali vines. With all of the supplies (zip lock bags, scissors, silica gel bags, filter paper folded inside letter-sized envelopes, a writing board, some paper and sharp pens for recording) tightly packed, we boarded the ferry to Hvar.

The wind and rain made it difficult to remain at the ferry deck and observe the stone walls of Split's Diocletian's palace (3rd ct. A.D.) and the palm trees in front of it slowly disappearing in the distance. As we later found out, Emperor Diocletian (Caius Aurelius Vaerius Diocletianus) built the palace in Split (Spalato) when he was at the height of his power at the turn of the 3rd ct. The spacious palace that today makes up the city center of Split has three temples, the most famous of which is the Temple of Jupiter. Diocletian dedicated the temple to the god Jupiter, because he considered himself to be his earthly son. The entrance to the temple is richly decorated with images of vine tendrils winding around the Erots endulging in grapes. In between the Erots, various animals (a dog, a cow and a goat) are shown eating the grapes. Historians believe that the decorations portraying vine-related activities, shown on the temple of Jupiter as well as inside Diocletian's mausoleum, are certain confirmations of the emperor's personal affinities towards winegrowing.

Upon arrival on the island, we ventured off to the Plavac mali vineyards. Everyone was anxious to take the first samples, but it seemed as if the weather was not

Reconstruction of the Diocletian's palace in Split (Source: Hrvoje Matijević, Museum of History Split).

Plavac mali vineyards on the island of Hvar - locality Sv. Nedilja (photo by Boris Kragić).

cooperating. Although Hvar is well known as the sunniest island in the Adriatic, with an average of 2720 sunny hours per year, the first rainy day did not promise much. Careful sampling and recording were almost impossible underneath the umbrellas, with the harsh wind blowing incessantly. Tired and hungry, but nevertheless under the strong impression of Hvar's jungle-like beauty, everyone agreed to a dinner in a local restaurant in the town of Hvar.

While walking the streets, an eerie feeling overcame us. It felt as though time had stopped in the heart of this little town, and took us back to ancient Pharos. The slippery stone paving of the streets in Hvar reminds the visitors that it has witnessed the passage of centuries and civilizations, overcoming all hardships with its quite presence. The architecture of Hvar is typically Mediterranean, with elements inherited from the Greeks and Romans. Stones are built upon older stones, in certain places replaced by new stones. Only the wooden, slightly carved window openings, along with the sounds of a stereo player, hinted the presence of modern inhabitants. A labyrinth of narrow streets led to our place of destination. The entrance to the restaurant was almost invisible from the street; the lights in the hallway were already dimmed, and a familiar smell of grilled fish spread outside the kitchen. Wooden tables, wooden chairs, the reminders of "a living with the sea" were everywhere around. A long fisherman's net with tangled seashells decorated the main wall, in a semi-circle shaped dining room. This family restaurant, owned by two brothers, turned into a hotspot of a very interesting dispute. Upon learning that their dinner guest is a renowned California grape geneticist, the brothers decided to let Meredith judge whose wine is better. One after the other, the glasses circled the table, and two bottles of Plavac mali were brought out. As if their lives depended on it, these two men focused their eyes on the professor's face, after she took a sip of one, then the other. The suspense was evident and an honest opinion that both wines are good, but that one should be left in the barrel to age a bit longer was quite satisfying to the hosts. A lively dinner discussion continued, with the serving of Dalmatian seafood specialties; grilled fish (orada and brancin, the highest quality white fish), and sautéed scampi and seashells in rice. The dinner ended with the famous dubrovačka rozata (a flan cake with caramel aroma) and Prošek, an excellent white dessert wine made from the grapes of Pošip, an autochthonous wine grape of neighboring Korčula.

The next day, we continued the search under more favorable conditions - the clouds cleared up and, all of a sudden, the island glowed in all its beauty. The locals say that at this time of the year, Hvar is the most impressive. The landscape bursts with yellow, purple and green colors, as the narcotic smell of lavender, laurel, and immortelles spreads through the air. Starting early in the morning, with the help of a local winegrower, Ivan Carić, we visited 10 vineyards and sampled a total of 31 Plavac mali vines. Upon sampling the vines in Ivan Dolac, a small village located at the southern end of the island, Carić insisted that everyone take a walk over to the church. He pointed to the stone plaque above the entrance doors dating back to 1901. The letters were still recognizable and, in translation, the writing read:

"In the name of the mother of God and her glory, this church was built by Ivan Carić, the son of late Juraj. Since 1852, oidium and peronospera have been on a

The plaque on the church of St. Rosary in Ivan Dolac on the island of Hvar - a prayer to the Blessed Virgin Mary (photo by Boris Kragić).

rampage through our vineyards. There were great troubles. Phylloxera has reached Zadar. Our vines are dying. We await the disappearance of our nation. People! In the name of your faith, turn to the blessed Virgin Mary for help and may God keep you safe from these three evils."

The year below, 1901, Mr. Carić continued with his explanation, is when Dalmatian viticulture experienced the culmination of its hardships. Oidium, peronospera and phylloxera raided the vineyards in Dalmatia, completely ruining more than 20% of them. Only the vines planted on remote islands, like Vis, remained somewhat spared.

Back to the Hvar vineyards, in the middle of an exhausting day, Pejić and Maletić suggested that we pay a visit to the renowned and most highly acclaimed winemaker on the island, Zlatan Plenković. Plenković took us on a tour of his winery and the neighboring vineyards,

Two top quality wines produced by Zlatan Plenković's winery in Sveta Nedilja on the island of Hvar (Source: Several authors - Hrvatska vina i vinari, Zagreb 2002).

most of them, of course, Plavac mali, the predominant red wine variety on the islands of Hvar, Korčula, Brač and the Pelješac peninsula. His winery, located at the tip of the island in a small village of Sveta Nedilja is a family business with a long tradition. That year, Plenković's wine Zlatan Plavac, made entirely from Plavac mali grapes, won the gold medal at the national exhibition of wines held each year in Zagreb. Plenković had a reason to be proud of his work. His assortment of wines included the dark ruby red Zlatan Plavac with 12.5-14.5 vol % alcohol, Zlatan Rosé with 12.0-13.5 vol % alcohol and Zlatan Otok, a high quality white wine made from the grapes of Bogdanuša, Prč and Mekuja, with 12-12.5 vol % alcohol. The vineyards beneath the stony cliffs, with perfectly aligned vines on the sloped terrain surrounding the winery, were indeed memorable. We left Sveta Nedilja with a dozen supreme quality Plavac wines, a gift from Zlatan Plenković.

The day ended, with the first Plavac mali leaf samples neatly packed inside the envelopes. Between 5 and 10 young leaves (about 0.4 inches in diameter) were placed between folded sides of a filter paper inside envelopes containing dessicant to speed up the drying process. This way, by the time the bags reached Davis, the leaves would be dry and ready for processing.

It was time to move on. Everyone was sorry to leave the lavender-packed fields of Hvar, but at the same time, eager to explore both Pelješac and Korčula. At the end of the ferry ride, the sun slowly dropped somewhere behind the flag on the Marjan hill (the viewpoint at the edge of Split from which most of the city can be seen). The Zagreb scientists wished us good luck and headed back home, while the road to Pelješac and a four hour drive back to Grgich's winery in Trstenik were still ahead of us.

We left the Split port and headed for Pelješac, after receiving short instructions from my friends Maja and Damir. Trying to hint the importance of our visit, I partially relieved my valuable wine cargo by leaving two bottles of Zlatan Plavac with my friends in Split. Although we were both tired from the vineyard hopping tour, Professor Meredith still stopped by the road on several occasions to take pictures of the sunset (Jadranska magistrala - the main road from Split do Dubrovnik is very similar to Highway 1 in California; it runs next to the sea for most of its length). The road was clear, and when I was just about to fall asleep, the lights of a police car driving behind us interrupted the idyll. I glanced at the speedometer and realized that I hadn't warned my boss about the speed limits in Croatia. On a moment's notice, we were forced to pull over and explain our intentions of quickly reaching Trstenik. Professor Meredith believed that the speed was appropriate for the road conditions, but the young policeman obviously felt that we were endangering his authority, and invited us to take a look at the radar on his car. He requested documentation and a driver's licence. Since he did not understand English, I took the papers and stepped outside. Just as in the case of many critical situations, my instinct came through. I figured that a California ID, or in this case, a driver's license, might impress the authorities, just as it did me a few years ago when I passed the driver's test in Davis. I showed the professor's license to the young policeman and added that my boss is on an important research mission that might benefit our country. I mentioned that everyone is helping us on this important task and that we would appreciate his benevolence. The young man inspected the license and, obviously impressed, smiled, returned to the car with a soundly pronounced "good luck". We waived good bye, and hoped that he would not change his mind a few miles down the road.

After our adventurous ride, we reached Trstenik shortly after midnight. There was still a lot of work ahead of us. The Plavac mali vineyards on both Pelješac and Korčula awaited our inspection.

Pelješac – the home of highest quality Plavac mali wine

Next on the repertoire were the vineyards on Pelješac; our goal was to visit as many vineyards at different locations and take several representative samples from each one. Professor Petar Maleš, the former head of the Institute for Adriatic Crops in Split did an extensive study of the Plavac mali variety, compiled in the publication "Population Plavac". According to his research results and morphological differences observed between the vines, Plavac mali was supposedly a heterogeneous variety with many different subtypes (such as Plavac mali gray, Plavac mali round etc.). Thus, due to many different localities on the 40 miles (65 km) long Pelješac peninsula, there was a lot of work to be done.

The Postup locality on the Pelješac peninsula - Plavac mali vineyards (photo by Boris Kragić).

Traveling through vineyards, one could not help but notice the diligence of the local field workers. In many places, the soil is sparser than the stone. The vineyards slope toward the sea at almost a 90° angle (Dingač vineyards). Enormous amounts of patience and care have to be invested when working around the vines in Postup and Dingač because this means working with hands and hands alone. The mechanization is too heavy and bulky to be taken even close to most of the vines in this region; the only help that people can rely on are donkeys.

In the next few days we made frequent stops in numerous vineyard regions of Pelješac: Dingač, Postup, Potomje, Kuna, etc. In total, we sampled 105 Plavac mali vines in 28 vineyards on the Pelješac peninsula.

Enchanting vineyards of Korčula

The last on the list was the island of Korčula. A ferry ride from Orebić (located at the tip of the Pelješac peninsula) to Korčula, took less than half an hour. With the help of a viticultural expert from Dubrovnik, Ivan Kiridžija, we visited 28 vineyards and sampled 105 Plavac mali vines on this beautiful island.

Throughout the centuries, interrupted by numerous wars and foreign colonizations, the inhabitants of Korčula have cherished their customs in viticulture and wine production. It is believed that the Greeks from Issa first introduced the vine to

The terraced vineyards carved in stone - a common site on the island of Korčula (photo by Boris Kragić).

the reddish sandy soil of Lumbarda on Korčula. The winegrowing regions later spread across the entire island and reached the surroundings of nowadays villages of Blato, Čara, Smokvica, and Žrnovo. The pride of the entire island is the winegrowing region of Čara-Smokvica, where special attention is paid to the white variety Pošip. Together with Plavac mali, Pošip and another autochthonous white cultivar, Grk, render supreme quality wines on Korčula.

The typical vineyards on Korčula are terraces, most of them facing the sunny southern side of the island. Unfortunately, many of these vineyards have been uprooted and abandoned in the past century, partly due to the trend of young people flocking to the cities, and partly due to inadequate governmental incentives allocated for restoration of viticulture devastated by diseases and pests. We learned that in the past 50 years, Korčula had one of the highest emigration rates of all Dalmatian islands. As a consequence, what remains today of Korčula's vast vineyards are dry-stone walls witnessing their not-so-recent fruitful existence. The vineyards on the island are best viewed from the air, since most terraces are located on sloped, hardly approachable terrain.

Just like Hvar, Korčula has a rich history extending back to the antic period. In the time of the Greeks, Korkyra (4th ct. B.C.) was a famous fort, and, more recently, many locals even argue, the home of Marco Pollo. Several historians have looked into the lineage of the Pollo (de Pollo) family and were able to trace close relatives back to Korčula. The streets of Korčula today peacefully hide their secrets, and the legends are retold from one generation to the next, while their verification is left up to the historians. We leave this town and the island, with the research mission accomplished.

Our last day together in Dubrovnik

After taking one last long look at the clear-blue sea in Trstenik from the balcony in my room and bidding a farewell to Mike Grgich, we headed for Dubrovnik. Professor Meredith had already gotten accustomed to the winding main road - the beautiful scenery that followed us throughout the ride made our trip to the far southern tip of Croatia really enjoyable. A few days earlier at the dinner we had at Jozo's house, I asked his granddaughter Ivana to be our guide in Dubrovnik. After our arrival, we had just about enough time to take a brief sightseeing tour of this breathtaking medieval town.

The Lovrijenac fortress of Dubrovnik located on the outskirts of the city walls
(photo by Jasenka Piljac).

The inhabitants of Dubrovnik take special pride in the cultural and historical heritage of their town. In describing Dubrovnik, they always mention two things, its enlistment in UNESCO World Heritage List and a quote by Bernard Shaw: "Those who seek paradise on earth should come and see Dubrovnik". More recently, as Ivana pointed out, Dubrovnik is also known as one of the cities hardest hit in the recent aggression of the Yugoslav army on Croatia (1991-1995). The entrance to the old city shows a map of the city center covered with black dots-locations where shells directly fell on this "pearl of the Adriatic". Right after the entrance through the main gate, we agreed with Ivana's suggestion to circle the two-kilometer long city walls surrounding the stone-paved nucleus.

Although Dubrovnik was founded in the 7th century, the stone walls were not built until much later, in the long period between the 11th and 17th centuries. Their main purpose was to offer protection against foreign invaders during the establishment of Dubrovnik as a city-state and, later on, the Republic. Today, along with the central street, Stradun, the city walls are the main tourist attraction. Indeed, the beauty of Dubrovnik is most impressive from a bird's eye view.

After circling the entire length of the walls, we decided to take a stroll through the city center, via Stradun. Ivana pointed out that Dubrovnik harbors one of the three oldest pharmacies in all of Europe (14th), located on the premises of the old Franciscan monastery. We took a glance at the Cathedral, St. Blaise's church (the saint protector of Dubrovnik), the Rector's palace and the Synagogue (also one of the oldest in Europe). One can easily find one's way in Dubrovnik, because most of the smaller streets eventually turn back to the main walkway of Stradun. As we later learned, one of the streets parallel with Stradun, called the Ulica Prijeko, is the home of the oldest vine in the region. Believed to be Malvasia dubrovačka, this about 200 year old vine emerges from the stone-paving and climbs several meters high up to one of the old stone balconies in the middle of the street. It still bears fruit each fall, and provides a unique decoration to its owner.

After completing our sightseeing tour and taking numerous pictures in Dubrovnik, we thanked Ivana for the hospitality in her hometown and left for the hotel Croatia in Cavtat where we were to spend the last night of our stay in Dalmatia. At its southernmost end where our hotel was located, Cavtat is thrust out on a rim of the coastal line and protrudes into the sea. The view of Dubrovnik from the hotel terrace reminded me of the postcard images we saw on Stradun – I concluded that many photographers had to go to Cavtat or take pictures from the air in order to get a view of the entire city of Dubrovnik. At nighttime, the illuminated stone walls surrounding Dubrovnik and its fortresses appear as ghostly images and reminders of the time and people gone by, arousing the feelings of respect and admiration. With this unforgettable image in my mind, I fell asleep to the sounds of the sea beneath my window.

The next morning we left our rental car at the airport and boarded the plane to Zagreb. We were both content with the results of this short expedition, and with *the variety of collected Plavac mali samples. I was happy for having had the opportunity to participate in practical science alongside my mentor, in an informal atmosphere of beautiful Dalmatian vineyards. We said our good-byes at the Zagreb airport, from where Professor Meredith continued on to California with more than 100 original Croatian souvenirs packed in her bags.*

ಲಿಲಿಲಿ

Professor Meredith took the dehydrated leaves collected from the 148 Plavac mali vines back to her lab in Davis, where DNA profiles were obtained. In addition, several Plavac mali samples from various sources were analyzed, among them Plavac mali – French, a mislabeled vine from the collection in Montpellier (France)* that was subsequently matched with the red wine variety Dobričić. It is assumed that this last sample was sent to France in 1951 by Marcel Jelaska from Split, and that misnaming might have occurred due to the synonym "Šoltanski Plavac" used for this cultivar on the island of Šolta.

Out of the 148 sampled vines from vineyards located on Hvar, Korčula and Pelješac, 136 vines were genetically identical. For the remaining 12 vines, it was not possible to precisely determine the genetic profile, or some discrepancies were observed which could point to variability within one variety or contributions from other genotypes. One of the 12 genotypes was identified as a separate variety, Plavina. This discovery indicated a mistake during sampling, which is easy to make if sampling is done at the time of early vegetation in May. The finding that 136 Plavac mali vines are genetical synonyms pointed to the genetic uniformity of this variety and was contrary to our initial hypothesis about genetic variability of Plavac mali that should have accounted for different types of Plavac (small, large, gray, etc.).[19] However, since the 1998 study was not undertaken with the purpose of discovering genetic variability within a variety, further analysis is needed to confirm this conclusion.

It was expected that at least some of the sampled Plavac mali cultivars would correspond to Zinfandel, but the analysis showed no exact match. Only a high degree of relatedness was established between Zinfandel and Plavac mali, pointing to a possible parent/progeny relationship between the two. Also, two other native Dalmatian varieties, Plavina and Grk, were confirmed as possible relatives of Zinfandel and Plavac mali. The trail was growing warmer, but, to the great disappointment of Mike Grgich, Plavac mali definitely wasn't the Croatian counterpart of Zinfandel. If one takes a closer look at the viticultural characteristics of Zinfandel and Plavac mali, as well as their growing regions, one can't help but notice that the major viticultural indicators confirm this fact. Professor Nikola Mirošević, the head of the Viticulture and enology department at the Faculty of Agriculture of the University of Zagreb, performed a parallel planting experiment with Plavac mali, Zinfandel, and Primitivo in his vineyard in Blato, Korčula. Even

---

* Grape collection INRA - Domaine de Vassal, 34340 Marseillan, France. This accession was obtained from Split, Croatia on March 13, 1951 and corresponds to cultivar Dobričić.

Limestone-rich regions are especially suitable for cultivation of resistant varieties like Plavac mali. This photograph was taken in an old vineyard in Kostanje, the hinterland of Split (photo by Jasenka Piljac).

before the DNA analyses were complete, Professor Mirošević pointed to the differences in the stages of development between Plavac mali on the one side and Zinfandel and Primitivo on the other.

The growing regions of Plavac mali (middle and southern Dalmatia), Primitivo (Italian viticultural region of Puglia) and Zinfandel (the California regions of Napa, Sonoma and Central Valley) do not belong to the same climatic belt.[26] Based on the climatic and ecological conditions of growth, Puglia and southern Dalmatia belong to the fourth climatic region according to Winkler, with the heat summation temperature (ΣEt — the sum of the mean monthly temperature above 10°C for the period concerned[16]) greater than 2205 °C. Napa, Sonoma and the Central Valley of California encompass several climatic regions. Napa and Sonoma extend between the second and fourth climatic regions, while Central Valley falls between the fourth and fifth regions. An appreciable yield for Plavac mali may be expected exclusively in the Croatian fifth climatic region (which corresponds to Winkler's fourth), at especially favorable *terroirs* that are exposed to direct and reflected sun rays, such as the southern points of Pelješac, Korčula, Hvar, Brač, Kaštela, as well as Mosor and Biokovo foothills. The effective heat summation temperatures determined for these locations significantly exceed 2205 °C and amount to: ΣEt °C: Split - 2248; Dubrovnik - 2253; Hvar – 2278; Opuzen – 2317. At most localities rich in limestone, the heat summation temperatures are not as high and, as a consequence, Plavac mali yields lower quality fruit. This is expected for all cultivars that ripen in the fourth ripening decade. Thus, Plavac mali is not distributed over a large geographic area, its northern coastal border ends in Primošten.

At all mentioned *terroirs*, Plavac mali reaches technological ripeness by the end of September at the earliest, and regularly, in early October.

According to Winkler,[16] Zinfandel is best adapted to the coastal valleys belonging to the second and third climatic regions where it gives superior quality dry wine precisely because of cooler growing conditions. In region IV, it is made into excellent dessert wine, "porto". Under these climatic conditions Zinfandel ripens right after Chardonnay, which corresponds to the second and early third ripening decades. The same applies to Primitivo grown in the Puglia region of Italy (IV. climatic region) where this variety ripens in late August and the first decade of September.

From the climatic and ecological analyses performed for Zinfandel in California and Primitivo in Italy,[26] it became clear that these cultivars belong to the II. and early III. ripening periods, based on the optimal heat summation temperature range from 1371 °C to 2204 °C. However, Plavac mali is a very late variety that belongs to the IV. ripening period and gives optimal yield only in the IV. climatic region according to Winkler, at southern *terroirs* exposed to direct sunlight, where ΣEt exceeds 2205 °C. Theoretically, Plavac mali could ripen under climatic conditions of the Puglia region, but much later than Primitivo, and it could never reach maturity in the II. and III. climatic regions of California, which in Croatia correspond to Slavonia, Podunavlje, Istria, the northern Croatian coastal belt, and Dalmatian Zagora. In accordance with the DNA analyses, it became clear that Plavac mali could not be the same variety as Californian Zinfandel.

Slightly puzzled with the first "search of the territory" in 1998, but nonetheless content with the work performed, Professor Meredith decided to let her new Zagreb colleagues, Pejić and Maletić, continue the search after her initiative. I remained in contact with both sides and waited for another chance to again become involved in the project that has become much more than a simple scientific investigation. Since, at the time, the plant genetics laboratory at the Faculty of Agronomy in Zagreb wasn't equipped to perform the expensive DNA analyses, I hoped for another return to the Meredith lab.

My wish was granted in the spring of 2000, when the naturalization process required me to take a trip back to Davis and accidentally happened to coincide with the time when Pejić and Maletić collected another interesting package of "Zinfandel suspects". I booked my plane ticket in April, but since my brother Ante was scheduled to visit our parents in Davis a few weeks earlier, in March, he had the priviledge of taking "the bag of suspects" across the border to California. By this time, everyone in my family was familiar with the search for Zinfandel, including my brother. He later told me that he boarded the plane with the valuable cargo, wondering which little plastic tube might contain the DNA that the entire California winemaking world was after.

Searching the vineyards of native Croatian varieties

Back in the lab, upon unraveling the papers containing cultivar descriptions, Professor Meredith found out that all of the additional sampled cultivars (12) sent by Pejić and Maletić come from southern Dalmatia, Split and the neighboring islands, where Zinfandel's close relatives, Plavina, Plavac mali and Grk were already detected. Because it is now known that many wine cultivars have arisen from natural crosses, particular interest was taken in cultivars with physiologically female flowers, such as Grk, Bratkovina b. and Cetinka, and cultivars such as Plavac mali and Plavina that could have served as their pollinators. Scientists from Zagreb hypothesized that the island of Korčula might have been a particularly rich source for the emergence of new cultivars because it has a favorable climate for grapevines, is geographically isolated, has a long winemaking tradition (written evidence from 4th century B.C.),[27] and was historically an important crossroads between Greece, Rome and Venice. The above-mentioned cultivars have been grown on Korčula for centuries.

Plavac mali, Zinfandel and the 12 new varieties from Croatia (Grk, Plavina, Vranac, Cetinka, Pošip b., Bratkovina b., Bratkovina c., Zlatarica blatska b., Dobričić, Brački crljenak, Vugava b., Babica) were selected for the experiments in the 2000 study because of their origin, similar phenotype, and flower and leaf appearance resembling Plavac mali and Zinfandel. In addition, the available DNA from the UC Davis database for the varieties Zinfandel, Plavac mali, Vranac, Plavina and Grk was also included in the experiment, as a positive control of the experimental procedure. Since one of the new samples, according to the "passport data" provided by the scientists from Zagreb, was a strong Zinfandel candidate, in a very superstitious manner, I took special care of that one from the beginning. The rest of the samples also had tags, saying: "This one reminds us of Zin, might be a relative" or "We are not sure what this one is".

In the experimental process, it was often hard to make sense of the final results of the analyses – the "numbers" representing DNA fragment sizes that defined the particular cultivar, separating it from the rest. My family members were often dragged into after-work discussions and eagerly monitored the progress on the "Zinfandel trail". At one point, I had computer printouts of the DNA profiles for each of the varieties, spread out all across the floor in my living room. Soon afterwards, I realized that the comparison of DNA profiles has became too cumbersome to be done by hand. Gerald Dangl, my supervisor and friend, who always looked at inventing new techniques to simplify our work, convinced me that a DNA program developed by one of my lab colleagues could do the work as correctly and at the same time make my job easier. I still checked many of them by hand, especially when a potential Zinfandel match came up. The entire laboratory team spent hours of scientific fun and suspense, attached to the lab computer at the end of Wickson hall.

In the experiments in 2000, I analyzed all 14 varieties at 25 DNA microsatellite marker loci and compared the data to over 300 cultivars in the UC Davis database. We concluded that all 12 varieties are indeed unique but that, once again, to our disappointment, none of the Croatian varieties was a perfect match for Zinfandel. A simple inspection of the DNA profiles again pointed out that Zinfandel shares many markers with the tested varieties from southern Dalmatia. In order to search for possible parent/progeny relationships among the varieties, computer analysis was once again employed.

Plavac mali – the child of Zinfandel

The search for parental relationships performed within the group of analyzed cultivars as well as within the entire UC Davis database containing more than 300 profiled cultivars resulted in a discovery that came as a surprise to everyone: **Zinfandel** (hermaphrodite) together with a native Dalmatian variety **Dobričić** (hermaphrodite) is the parent of **Plavac mali** (hermaphrodite)! Of 14 closely related cultivars analyzed, no other pair could be a better parental fit, the nearest possibilities being excluded with preliminary testing. Furthermore, a computer-assisted search for parents in the existing UC Davis database with over 300 profiled cultivars from all over the world resulted in no other possible parental combination for Plavac mali!

Cultivar Dobričić is today grown almost exclusively on the island of Šolta, where it has a very long tradition.[21] In the past, up until 1929, Dobričić was widespread on the islands of Šolta and Čiovo, in the Split surroundings (Donji Kašteli), in Brusje on the island of Hvar and Milna on the islands of Brač. Today, besides on Šolta, individual vines may be found in the vineyards on the neighboring islands of Brač and Čiovo (where it is known under the synonym Slatinjanac) and at certain locations in the Split surroundings. Experts estimate that only about 250 acres (100 ha) of Dobričić are planted on Šolta and warn that this variety is seriously endangered.[28]

Professor Carole Meredith and Jasenka Piljac at the 2000 conference, Prospects for viticulture and enology, held in Zagreb. The preliminary results of the analysis of the relatedness between Zinfandel and autochthonous Croatian wine grape varieties were presented in the form of posters (photo by Snježana Mihaljević).

Dobričić grapes yield deep red wine whose quality was highly acclaimed in the past. In order to achieve color improvement, the wine was often mixed with the other red varieties. Also, the bitter taste of the wine and thick texture made it suitable for mixing with other, lighter versions of Dalmatian reds. Since the cultivation of Dobričić was restricted to the mentioned locations in the past, it is reasonable to assume that the cross between Zinfandel and Dobričić might have occurred at the island of Šolta or the nearby surroundings (the islands of Čiovo and Brač), where this cultivar has the longest tradition. This suggests that Zinfandel was once grown in this area as well.

Knowing the parentage of cultivars is not only of historical interest, it is important in the reconstruction of breeding events and in studies of relatedness of major grape varieties. This information can be invaluable in predicting the outcomes of future deliberate crosses and discovering the true lineage of classic wine grapes grown today. The pairing of Plavac mali, Dobričić and Zinfandel in a family relationship came as a surprise in the sense that this discovery points to Zinfandel as an even older cultivar than Plavac mali, which has been grown in Dalmatia for at least 150 years.

Relatedness of Zinfandel to other Croatian cultivars

Based on the available genetic data of certain autochthonous Croatian varieties,[29,30,31] we constructed a diagram showing putative relationships between Zinfandel and Croatian cultivars. It is evident that Zinfandel is closely related to many cultivars believed to be native to Dalmatia, such as Grk, Vranac, and Plavina and that it might be a distant relative of many others. Grk, Vranac and Plavina have a long tradition in South Dalmatia,[21,32] which provides additional evidence for the Dalmatian origin of Zinfandel. Moreover, in the past it was probably widely grown and played an important role as the pollinator in the development of new cultivars.

Zinfandel is in the center of complex relationships involving mostly Croatian cultivars. A valid hypothesis is that Zinfandel was an "old settler", who contributed to the Croatian *Vitis vinifera* L. gene pool by serving as a pollinator (in intentional breeding events or unintentional spontaneous crosses) of the local varieties grown in southern Dalmatia and on the islands. It is highly probable that the cultivation of Zinfandel was discontinued in the past, due to pests or diseases (especially phylloxera) or even wars and trade. It gradually became replaced by a biologically more resistant variety that met the demands of local winegrowers; namely, its offspring, Plavac mali.

Chances of finding Zinfandel in other countries

For a more complete analysis of Zinfandel's origin, I examined the chances of finding Zinfandel in Italian, Greek, and Croatian *Vitis vinifera* populations.[29] In addition to our focus on Croatia, it was necessary to eliminate, on a scientific bases, the chance that Zinfandel might have originated in one of the neighboring

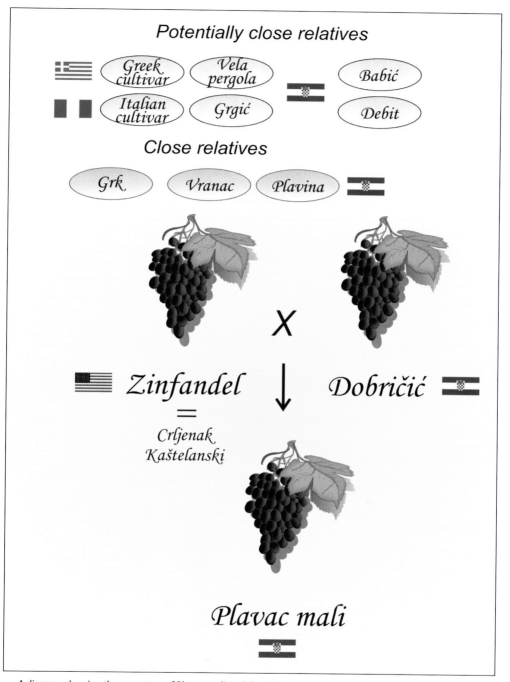

A diagram showing the parentage of Plavac mali and the relatedness of Zinfandel to autochthonous Croatian varieties. The diagram was constructed based on DNA microsatellite data, whereby blue color indicates a high degree of relatedness and magenta points out the more distant relatives (design by Ante Jukić).

countries that also had a long tradition in viticulture. The genetic profiles of Greek and Italian varieties were already published and readily available on the internet (htttp://www.boku.ac.at/zag /forsch/grapeSSR2.htm). The analysis employed 32 Greek, 30 Italian, and an additional 15 Croatian cultivars (Babić, Debit, Gegić, Lasina, Bogdanuša, Pošip n., Žilavka, Hrvatica, Teran, Vela pergola, Žlahtina, Malvazija istarska, Kraljevina, Škrlet, Ranfol),[30] comprising a set of 30 Croatian cultivars– primarily from coastal Croatia. The results of the analysis, based on the relative frequencies of specific repetitive DNA sequences, indicated that Zinfandel is most likely to be found in the Croatian grapevine gene pool.

With the available genetic data, the probability of finding Zinfandel in a subgroup of Croatian cultivars, as opposed to a population of cultivars comprising the entire Meredith database at UC Davis, was also examined. The database contained more than 300 profiled varieties grown all over the world. More specifically, this time the comparison was targeted at identifying rare sequences of DNA found only in Zinfandel and Croatian cultivars. A pair of such rare DNA sequences was discovered to appear only in Zinfandel and the Croatian cultivars at an appreciable frequency.

Given the amassed scientific evidence, and likelihood ratios that added statistical weight to our findings, I felt content with the conclusion that Zinfandel has its roots in Dalmatia where today, its numerous relatives are continuing the "legacy" of their famous ancestor.

Kaštel Novi – the home of Crljenak and Zinfandel

Although the genetic analyses were sufficient to proclaim Dalmatia as the original home of Zinfandel, the drive to find original Croatian Zinfandel did not subside. Pejić and Maletić continued to search old vineyards in the surroundings of Split. They received great help from the knowledgeable local grape grower and manager of the Kaštelacoop wine cooperative, Ante Vuletin. The "suspect" Zinfandels were selected either in old vineyards, or vineyards whose owners used to collect native varieties. Disappointed to hear that the best Zinfandel candidate did not live up to their expectations, Pejić and Maletić decided to resample the "hot candidate" from 2000. In 2001, they went back to the same old vineyard recommended by Vuletin, located in Kaštel Novi, a coastal town north from Split. There they realized that, because of vigorous growth, the branches and leaves have intertwined themselves with the neighboring Babica vine, and that the candidate from 2000 was wrongly sampled. After a six-year-long search the persistence finally paid off in 2001, just before Christmas. In her lab in Davis, Dr. Meredith again analyzed this "suspect" Zinfandel, a vine locally known as Crljenak kaštelanski (meaning, "the red from the town of Kašteli"). This time, the DNA profile exactly matched Zinfandel! Professor Meredith sent e-mails from Davis, saying: "We have a match for Zinfandel. Quite convincing, finally!" An additional 40 vines from that vineyard, all resembling Crljenak, were subsequently also analyzed and nine additional ones matched Zinfandel. I was thrilled to have made a contribution to such an important discovery, as the newspapers and wine magazines of California reported that the Zinfandel mystery had finally been solved!

The first official presentation of the research findings occurred at the 2002 Zinposium held on June 15, in Santa Rosa, California.[33] The American-Croatian team of scientists (Professor Carole Meredith, Professor James Wolpert and Drs. Ivan Pejić and Edi Maletić), including Professor Charles Sullivan, had finally assembled in California to reconstruct the trail of Zinfandel, from its early beginnings in Dalmatia to its California-born prosperity and glory. In the enology session of the symposium, parallel tastings of Zinfandel and Plavac mali wines were performed in order to explore the taste similarity warranted by their close genetic relationship. Based on the small sample, it was concluded that

One of nine vines of Crljenak kaštelanski, the Croatian counterpart of Zinfandel, found in the vineyard of Ivica Radunić in Kaštel Novi (photo by Jasenka Piljac).

the Zinfandel wines are slightly richer and deeper, probably owing to the harvest year that resulted in lower yields. Many California winegrowers who attended the 2002 Zinposium have expressed an interest in obtaining the original cuttings of Crljenak kaštelanski from Croatia and visiting the original vineyard where it was discovered. Adding to their interest was Professor Meredith's own fascination with the Dalmatian coast, which reminded her of California's "Big Sur" in certain places, sloping right down to the sea.

The vineyard that will go down in history

Although, on many occasions, the participants of the Zinfandel search felt disappointed and hopeless, in the end, their luck and persistence had finally brought them to the vineyard of Ivica Radunić in Kaštel Novi. Had the search started a few years later, most probably, it would have been too late, because vineyards get replanted and the owner would not have noticed anything special about his

Ivica Radunić and Jasenka Piljac photographed next to the Crljenak kaštelanski vine in the summer of 2003 (photo by Hrvoje Matijević).

vines.[34] In fact, Radunić claims that, in the past, three types of Crljenak — Crljenak ljutun, Crljenak brački and Crljenak kaštelanski — were quite common in the vineyards of the seven Kaštela (seven small towns named Štafilić, Kambelovac, Novi, Stari, Sućurac, Lukšić and Gomilica), near Split. Although it is not known whether the three types of Crljenak are genetically distinct, each one exhibits different characteristics. High acid content is typical for grapes of Crljenak ljutun ("ljut" in Croatian means "bitter"), while Crljenak kaštelanski ripens unevenly and much earlier than the other varieties grown in this region (15th of August as opposed to 1st of October).

The particular nine vines of Crljenak kaštelanski discovered in Radunić's vineyard, spread across 2.5 acres of land (1 ha), are the heritage of his family's old vineyard in Kaštel Novi from which cuttings were taken 35 years ago. In addition to the three types of Crljenak, Radunić and his father also transferred the cuttings of Babica, a high-yielding variety well adapted to the red-soiled fields of Kaštela. Considering the long-treasured tradition in viticulture, Radunić approximated that Crljenak ran through the generations of vineyards in his family for at least a hundred years. As a curiosity, I should point out that in 2000, one year before the discovery of Zinfandel in his "backyard", Radunić was awarded the first prize for exemplary maintenance of a large number of autochthonous varieties in the County of Split. While Ante Vuletin from Kaštelacoop, in cooperation with the Institute for Adriatic Crops in Split and the Faculty of Agronomy in Zagreb, regulates the production of Crljenak cuttings intended for overseas shipment to interested California winemakers, Radunić has a small project of his own underway. Alongside the 35-year-old vineyard, he plans the planting of a new vineyard intended mostly for Plavac mali, Maraština and, of course, Crljenak. He had prepared 200 rootstocks for this year's grafting of Crljenak and hopes to have his first varietal wine released in 2007. A larger planting of Crljenak (several thousand cuttings) is planned for the south-facing vineyards in Sveta Nedilja on the island of Hvar, where Zlatan Plenković, a renowned winemaker famous for his Plavac, has the necessary infrastructure and available land to undertake such a project.

Although he always had a market for his wine, Radunić noted quite an impact of Zinfandel's discovery on his popularity among the local winemakers. With a sudden increase in sales, his cellar has become too small to meet the local demand. Currently, he produces approximately 6,000 liters of wine, of which white varietal Maraština accounts for 650 L, and the remainder is split in a 1:2 ratio between rosé and the reds (Crljenak, Babica). Unlike many high-quantity producers, Radunić always emphasized the quality of his wine and never considered dropping the standards in exchange for greater quantity. On his recent visit to Radunić's small estate in Kaštel Novi, David Gates, the vice president of Ridge vineyards in California, encouraged Radunić to continue with wine production and concentrate on the reds. He believes that with a slight change in technology, high quality wines suitable for export might be the future of winemaking in Kaštela. In the coming years, Radunić plans on expanding his small winery, increasing the number of canes (his vineyards currently totals about 6000) and establishing a small business, with tastings and souvenirs available to tourists throughout the summer season.

On a curious note, Radunić often jokes about the discovery of Zinfandel in his vineyard. As a small child, his friends in Kaštela nick-named him Billy the Kid, and ever since, he is known by the name "Kid" in the local community. Radunić believes that serendipity brought Zinfandel from the former Wild West back to his vineyard. As an active fireman, Radunić leads a very hectic life in the summer months, when his duties call on him to protect and safeguard not only his own vineyard but the vineyards of the surrounding Kaštela where more Zinfandel vines might be waiting to get discovered.

In the past, Kaštel Novi and Kaštel Stari were well-known ports from which large amounts of wine, especially red, were exported to France in the mid-1800s. The need for Dalmatian exports became even more pronounced when phylloxera infected large vineyard areas across Europe and a new, larger port was built in Kaštel Novi whose capacity could handle greater quantities of wine. Today, the ports in Kaštel Novi and Kaštel Stari mostly serve the tourists and local fishermen, since a rapid decline in winegrowing affected the seven Kaštela in much the same way as the rest of Dalmatian towns.

Continuing the search

The search for additional Crljenak vines in Croatia continues and, since the original finding in Kaštel Novi, several other vines in the hinterland of the coastal town of Omiš have been identified as Zinfandel matches. In the local community, these vines are known under the name Pribidrag. During her last visit to Croatia, in the summer of 2002, Meredith visited the island of Šolta where Dobričić was found, and the island of Čiovo where she believes she saw some promising Zinfandel candidates.[34] Being able to compare the Crljenak, Pribidrag, Zinfandel and Primitivo growing conditions, including climate, soil and vineyard maintenance styles, will give powerful insight into the ways the winegrowers in different countries are managing this variety.

The canyon of the river Cetina in the hinterland of the coastal town of Omiš - this region is particularly suitable for vine cultivation (photo by Jasenka Piljac).

From the scientific point of view, knowing the origin of a certain variety is useful for finding the subtypes or clones that have developed in the past through slow changes in the genetic make-up. If the origin of a vine can be traced down to a certain region, then most of its variants will be distributed in the vicinity of that locale. Until now, scientists had no place to look for Zinfandel subtypes and could not compare them with the California Zinfandel. The variants that are potentially still hiding in Croatian vineyards might be interesting for their advantageous properties that often become expressed after prolonged vegetative propagation, such as disease resistance, high temperature tolerance, more intense berry color, etc.

The scientists from the University of Zagreb are hopeful that they might even be able to trace down the parents of Zinfandel. The exact path of this vine from Croatia to America is still debatable, although Professor Sullivan's claim that it was imported to the east coast from the Schönbrunn imperial collection in Vienna seems the most plausible. Professor Meredith[34] does not exclude the possibility mentioned by Burton Anderson, an American wine writer living in Italy, and Darrell Corti, a wine merchant from Sacramento, that the monks who emigrated from Croatia, in order to escape religious persecution in the eighteenth century, brought the grape with them to Italy. There it became known as Primitivo.

However, it is also possible that in one of their numerous migrations to California, in search of a better living after the vineyard devastation in the early 19th century, along with their customs and traditions, Dalmatians packed a few cuttings of Crljenak kaštelanski. One thing is certain, Zinfandel was born in Croatia a long time ago, when Dalmatia alone numbered more than 200 autochthonous varieties and represented the Mediterranean breeding ground for wine grapes. On the fertile Californian grounds it flourished and eventually became a classical American success story - a symbol of not only the Californian wine industry, but the Land of Opportunities itself.

A wine road across the Atlantic

A famous old saying claims that "Water separates, and wine unites the nations". Owing to Zinfandel, the town of Kaštela has witnessed the uniting power of wine that is so often stressed by *connoisseurs*. On the 18th of October, 2002, on the same day when the town of Kaštela was awarded the prestigeous European award for the development and advances achieved in tourism, Mary Lou Holt, the mayor of Yountville, arrived in Kaštela. The objective of her trip was establishing a sisterhood between the Kaštela and Yountville towns. The major initiator of this idea was, once again, Mike Grgich. Grgich felt that, after five decades of winemaking experience marked by dedicated promotion of his native Croatia and its viticultural tradition in California, he managed to bring the two countries together through Zinfandel.

Yountville and Kaštela are similar in many ways, according to Holt, the added beauty of Kaštela is the Adriatic sea.[35] Yountville is situated in the heart of Napa Valley, the most prominent Californian viticultural region and is known for its long viticultural tradition, the center of which has been occupied by Zinfandel for a very long time. The winegrowing roots of Yountville extend back to 1838, when George Yount, its founder, planted the first vines in Napa Valley. This was the year that marked the beginning of the viticultural history of Napa Valley. The viticultural tradition of Kaštela and its surroundings extends even further back in time, but, just as many other Dalmatian towns, Kaštela experienced its rises and falls, and the continuity of winegrowing was often interrupted by natural and human factors, such as migrations, wars and industrialization.

In both towns, separated by hundreds of miles with the Atlantic in between, the inhabitants live similar peaceful lives in harmony with nature. Through the newly founded friendship between the two towns, the mayors, Mary Lou Holt and Ante Sanader, opened up a special kind of wine road, the one connecting Zinfandel on one continent and Crljenak kaštelanski on the other. According to Holt,[35] this road will give numerous opportunities for the two communities to build strong ties and learn from each other's experiences. Sanader sees the new bond between these small towns as a challenge for the people of Kaštela to turn back to viticulture and revive this old tradition in order to meet the new expectations set by their colleagues from Napa Valley. After all, the land in Kaštela is a God-given home for the vine.

The mayor of Yountville, Mary Lou Holt, and the mayor of Kaštela, Ante Sanader, holding a signed Charter of sisterhood confirming friendship between the two viticultural centers.
(Source: Mike Grgich).

CHAPTER THREE

VITICULTURAL TRADITION OF THE TROGIR-KAŠTELA SURROUNDINGS

The overall history of the Trogir-Kaštela region is closely tied to viticulture and winemaking, two main agricultural activities that ensured prosperity for its inhabitants for centuries. Wine was often the only means of survival and the farmer's destiny. Although the Kaštela region historically followed the viticultural destiny of Dalmatia, its geographical and climatic conditions are specific and the features of the wine produced there are unique.

The Trogir-Kaštela winegrowing region consists of hilly-plain motifs spread across the islands and inland. In this winegrowing region, the Kaštela field, referring to the plain beneath the hill Kozjak, is the most interesting. The history of

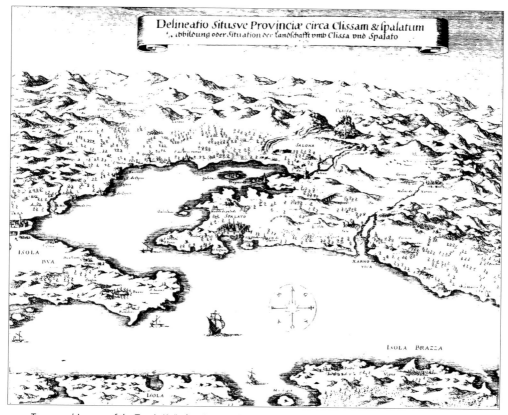

Topographic map of the Trogir-Kaštela winegrowing region in 1650 (Source: Archeological museum, Split).

this region is very rich with traces of the passing eras found everywhere around. The fertile ground and drinkable water of the Trogir-Kaštela region even attracted the Neanderthal caveman. The archeological findings from the so-called Mujina cave near Trogir (Trapljeni doci) confirm the presence of early tribal communities in the prehistorical period.[1] The tools uncovered in the cave, at least 50,000 years old, point to activities in this region at the dawn of humanity.

The Illyrian period

The members of the Illyrian tribe Delmati (Dalmati) inhabited the region of the Kaštela field. As far as the historical documents and the recordings of antic writers allow us to predict, initially, Delmati inhabited the inland region (today's Hercegovina), where the center of their tribe, Delminium, was located. In the 3rd ct. B.C. they migrated closer to the coast and settled between the banks of rivers the Krka and Cetina. About the same time when Delmati reached the Kaštela field, they noticed that the fertile plain was suitable for farming and cattle-breeding. Although Greek geographers mention the presence of several other tribal groups, namely, Manijce, Buline and Mile, Delmati certainly dominated all along the coastal belt. A large number of ceramics and stone relics dating back to the Neolithic period have been uncovered in the Kaštela field and all along the coast. They indicate that the Kaštela region has been inhabited since ancient times and the assumption of the Illyrians practicing agriculture there seems quite plausible.

Agriculture played a significant role in the every-day life of Delmats. The relics of primitive tools used in cultivation have been found in the remains of Illyrian settlements, indicating that the Illyrians not only gathered wild fruit but grew vine as well. In the Neolithic period, the most often used tool was a hoe made from antler. In the early iron period, iron hoes were used; however, they did not completely replace the tools made from animal bones. With time, iron hoes slowly adopted very functional shapes (the writer Hesihius mentions the *to oenotria* hoe used for the vine),[2] and those found in Dalmatia today do not significantly differ from the antic designs. The remaining agricultural tools used by Illyrians included, the shovel, the rake, the sickel and the scythe, while the plough was introduced later on during the settling of the Romans.

It is unquestionable that the Illyrians in Dalmatia were familiar with the vine in the bronze and early iron periods and that they produced wine before the coming of the Greeks. Although there is no concrete archeological evidence confirming an organized Illyrian wine production, this possibility cannot be excluded by any means. Since the investigations of Illyrian deposits have not been completed, a discovery confirming Illyrian vine cultivation in the Kaštela field may still be expected.

The Greeks in the Kaštela field

Of all the Greek colonies established on the shores of the Adriatic, the most significant was definitely Issa, whose colonists in the 2nd ct. B.C. established Tragurion (today Trogir), at the edge of the Kaštela field. It is certain that Tragurion was for a long time connected with its mother colony and that the Isseans carried over to Tragurion their knowledge and experience related to the vine. During the Greek period, the first vineyards appeared in the Kaštela field because its location, the composition of the soil, and insolation represented ideal conditions for winegrowing.

The remains of a helenistic port from the 1st and 2nd centuries B.C. were uncovered near Resnik in the Kaštela region.[3] Through this port, Tragurion and the surroundings were directly communicating with Issa and Pharos, as confirmed by the Issean silver coins found in archeological deposits near Trogir. Such a connection of the Kaštela region with the viticultural centers of Dalmatia significantly influenced the spread of viticultural knowledge. The evidence is found in antic ceramic pottery deposits uncovered near Resnik, containing numerous wine jugs and glasses as well as moulds used for their production. Although some historians still debate whether these findings point to the production or trade of wine, it is unquestionable that in the Greek period the Kaštelans lived in symbiosis with the vine.

The antic port near Resnik was undoubtedly the main center for the export of wine produced in the Kaštela field. This is confirmed by archeological deposits rich with amphoras and other wine-related pottery, found at the bottom of the sea near Split and the Kaštela bay. Although amphoras were used for the export of various food items, most of them were specifically intended for wine (*amphorae vinarie*).[4] Wine amphoras often had an elongated shape and were layered with a thin coating of wax on the inner side. The coating protected the inner side of the amphora from leakage and gave wine a specific noble pine aroma. According to certain sources, liquid wax obtained from conifers, especially pine, was used for preservation of wine. Amphoras of the Greek-Italic origin used for the transport of wine were usually made from a reddish-brown ceramic mass. The neck of the amphora was short with a tilted and finely shaped rim, and small, flat-shaped handles. The dimensions of amphoras did not exceed 36 inches (90 cm) in height.

In addition to amphoras, wine jugs (*oinohoe*) and wine glasses (*skyphos*) originating from southern Italy (Magna Graecia) were found along the Dalmatian coast.[5] One of the most significant collections of helenistic ceramics and pottery related to wine belongs to Kaštela and consists mostly of Megar glasses, named after a Greek city. In the antic world, they were especially valued and treasured during feasts. One specific item from the Kaštela collection is especially important – a very rare cup with a relief decoration of human and godly figures intertwined among the vine tendrils.

Most ceramic items discovered in the Adriatic belong to the 4th ct. B.C. and are decorated by white images painted on a black background. The most common motifs are birds and pigeons, although numerous wine jugs portraying owls, dolphins, octopus, as well as grapes, ivy and female faces were also discovered. It is assumed that many of the decorated wine jugs were produced in Issa. A represen-

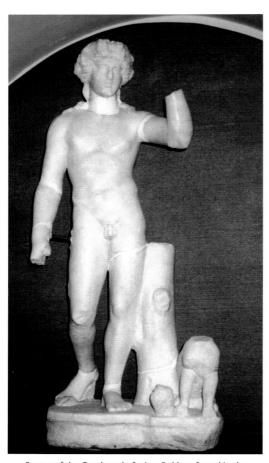

Statue of the Greek god of wine, Bakhos, found in the archeological deposits of the Kaštela field. The statue is kept at the Archeological museum in Split (photo by Jasenka Piljac).

tative statue of the Greek god of wine, Bakhos, was also found in the deposits of the Kaštela field. It is clear that wherever the wine god was worshiped, winemaking had to have been the center of attention. The overall life of Greeks in the Trogir-Kaštela region was dependent on the vine to such an extent that wine was often ascribed a cult-like meaning.

The Greeks transferred immense knowledge about the vine to the native Illyrians. Although, on many occasions, the newcomers and natives found common ground, problems also emerged between the fierce Delmati and Issean Tragurion. In one of their invasions, Delmati completely devastated the vineyards surrounding Tragurion.

It is believed that the ancient Greeks introduced to their Dalmatian colonies the knowledge of distinction between wine and table grapes. During the harvest, the Greeks separated the wine from table grapes and prepared them in different ways for later consumption. According to oral tradition, the custom in the Kaštela, whereby the raisins of early varieties are used to produce special wine types, also dates back to Greek times. On many occasions, in order to achieve faster drying than in the sun, grapes were immersed into basic solution.

The development of viticulture under Roman rule

In their aspirations to rule the entire Mediterranean, including the south-eastern shores of the Adriatic, the Romans made a pact with the Greeks in the 3rd ct. B.C., and became confronted with the Illyrian tribes. The conflict between the Romans and the Illyrians escalated and ended in a war in 229. B.C. that lasted almost 300 years. First, the Roman legions conquered the state of Ardieyans and later defeated the Ilyrian king Gentius. Long battles with three other Illyrian tribes — the Delmats, Japods and Liburnis — followed. The Roman province Ilyrika was divided into three legislative branches (convents) located in today's Skradin (*Scardoni*), Solin (*Salona*) and Vid near Metković (*Naroni*). The Kaštela region belonged to the Salona convent, while Trogir was a part of the Roman municipality joined to Salona – the capital of the Roman province Dalmatia (*Ilirikum*).

Because it represented the source of their power, the Romans treasured the vine and encouraged its expansion and wine production. Their tasks became easier upon reaching Dalmatia, where the vine had already established itself as the center of the economy. The newcomers continued to plant vineyards wherever the climate conditions allowed. Viticulture reached its high-point during the reign of Roman rulers because the Romans contributed their knowledge and continued to build upon the strong viticutural foundations of their predecessors, the Greeks and the Illyrians. It is thus justified to assume that the Romans worked on expanding the vineyards in the Kaštela field as well.

There exists no specific data on the type of vineyard maintenance practiced in the Trogir -Kaštela region; however, it is certain that the Romans highly regarded the vine in the Kaštela surroundings, just as in the case of their other winegrowing provinces.

The recent finding* of the remains of a well-preserved vine branch and a wine pot from the 2nd ct. AD, just across the sea from Kaštel Sućurac, attest to the presence of wine in this region during the Roman period. The antic writers Pliny, Katon, Horatius, Vergil etc. wrote a lot about the treatment of the vine in ancient Rome; however, Columella gave the most picturesque portrayal in his work "De re Rustica". In the 1st ct. B.C., Columella noted that the vine should be cultivated in dry and sunny locations on terrain facing the east and the south. The soil should be hoed up to the depth of 34-36 inches (85-90 cm). In the karst regions, the rocks had to be removed before hoeing, and ditches dug up for each individual cane. On especially dry terrain, the soil had to be removed all the way up to the depth of three feet (1 meter), and new vines planted at a distance between 2-4 feet (64 - 130 cm), depending upon the type of soil. Columella emphasized that the worst type of soil for a new vineyard was the one where vines were already grown in the past. Such soils had to be replenished by removing the old vines along with the roots, burning this old material and scattering the ashes across the entire area. In the beginning, vines were planted without paying attention to the cultivar properties, while the suitability of varieties to a certain growing region was later a major factor in cultivar choice. When the word spread that a specific variety gives good quality grapes, the locals swiftly turned to it. Writers from the antic period often recommended cultivars to winegrowers based on their adaptability to climatic and ecological factors.

The Romans dedicated special attention to winegrowing in Dalmatia; however, most of the secrets of their expertise remain unknown to this date.[6] It was customary to trim the surface roots and place bird fertilizer on top. Columella suggested monthly hoeing of the vineyard from March until October with obligatory

---

* This discovery was reported on Croatian national TV in October 2003, followed by an article published in *Slobodna Dalmacija* "What did the ancient peoples of Salona eat and drink?", by Meriene Jelača October 16th issue, 2003.

A portion of a sarcophagus recovered from the archeological remains of antic Salona (today Solin). The drawings portray a harvest somewhere between the year 240 and 250 A.D. (Source: Želimir Bašić - Dalmatinska vina kroz stoljeća, Šibenik 2001.).

removal of grass and weed. Because of successful agrotechnic measures, the Romans achieved relatively high yields - from 64 to 120 hl per acre (160 to 300 hl/ha). If the vineyard yielded below 15 hl, it was declared unprofitable and was subsequently removed. As far as vineyard style is concerned, fork-shaped vines were maintained by low growth. It is especially interesting to note that the handling of vines has not significantly changed in the vineyards of Kaštela and Dalmatia since the Roman times.

The fact that the Romans significantly improved the viticultural practices and increased the areas under vineyards with respect to the Greeks is supported by many archeological findings.[7] In the ruins of Salona, the former capital of the Dalmatian province, uncovered deposits revealed pieces of a wine press. It consisted of a rectangular slab with a round groove carved in the middle. Two indentations were most likely intended for the central beam on which two stones were placed for the necessary pressure. The squeezed grapes released the juice that flowed through the groove into a stone vessel. At the edge of the Kaštela field, near the archeological remains of antic Salona, a large portion of a sarcophagus was also uncovered, whose drawings portray a harvest somewhere between the year 240 and 250. Although mythical in its nature, the harvest is depicted in a quite realistic manner. The preserved pieces of the sarcophagus show a woman and the Erots (the small mythical gods of love, Aphrodite's companions) and parts of other figures that could not be identified. Between the vines and the Erots, a large horse carrying a basket full of grapes is easily recognized. The best preserved sarcophagus from the 3rd ct. A.D., also belonging to the Roman period, was found about

Roman sarcophagus from the 3rd ct. A.D., depicting Erots in harvest festivities. The sarcophagus was recovered in its entirety from the antic Salona graveyard twenty years ago (Source: Želimir Bašić - Dalmatinska vina kroz stoljeća, Šibenik 2001).

20 years ago in a Salona graveyard next to the road leading to Kaštel Sućurac. The sarcophagus was recovered in its entirety with the reliefs completely preserved. This murmur tombstone, completely decorated with vine leaves and grapes, depicts the Erots in harvest festivities. The sarcophagus also pictures various animals (goats, tigers, dogs, etc.) carrying baskets full of grapes on their backs. It is likely that this part of the relief is mythical, and does not represent the actual use of animals in harvest. However, the knives in the hands of six Erots completely resemble the knives used today for hand picking of grapes in Dalmatia. The relief found on this sarcophagus is especially significant from the viticultural aspect, because it gives a valuable insight into the ancient harvest customs in the Kaštela field during the Roman times.

Additional evidence confirming the existence of a wine trade,[8] such as stone-carved epitaphs mentioning the wine traders (*negotiator vinarius*), also abound in Salona. Good roads surrounding Salona, built by P. Kornelius Dolabela, played a significant role in the wine trade. They were used to transport wine from the fertile Kaštela field to the inland provinces. A solidly built road called Via Tragurina ran between the vineyards located in the fertile plains of the Kaštela field (*Ager salonitanus*) and connected Tragurion with the Roman capital, Salona.

The Roman soldiers, often unconsciously, played an important role in the spread of viticulture through the provinces. Along with their habit of wine drinking, they often brought cuttings of varieties to the newly conquered territories. Regarding Roman conquests, a famous saying states: "A Roman settles, wherever he achieves victory".[9] The Romans were aware that investing in agriculture, especially viticulture, would allow them to stay in power for years to come. Thus, in the first half of the 1st ct. the well-known Emperor Claudius brought his devoted followers and soldiers to the Siculi settlement located around the hill Bijaći, in the western part of the Kaštela field, and granted each one a piece of fertile land. Although no written records exist, it is highly likely that a large part of this land was used to plant vineyards.

Numerous buildings from the antic period, especially the *villae rusticae*, also attest to the significance of vine and wine in the every-day life of Romans. The remains of the *villae rusticae*, small houses built in the center of vineyard properties, were found in several places in the Kaštela field. It was customary for the colonists to found agricultural estates with the *villae rusticae* in the middle. They mostly served as small agricultural economies and as resting places for their owners. Based on archeological findings uncovered in close proximity to the *villae rusticae*, such as the fragments of monuments and items connected to the production of wine (amphoras, jugs and pots for wine, as well as wine presses), it may be concluded that their main purpose was served in the vineyards.

The remains of a large number of antic villae have been found throughout Dalmatia and in the Kaštela field, each surrounded by numerous deposits of stone pottery used for wine and oil. It is believed that in the antic period, the Kaštela field was a part of the Salona ager (field) that abounded in Mediterranean cultures, especially the vine.

The steps involved in wine production may be fairly well reconstructed with the help of numerous archeological deposits and historical documents.[10] It is certain that the process of wine production in Trogir and the Kaštela field was very similar to the one used on the Apenine peninsula. The treatment of grapes in Rome was described by numerous authors, especially Pliny, Katon, Columella, and Vergil.

Wine production started with the crushing of grapes in a special room called *torcularium*. From this room, the must was poured through the pipes made from baked clay

Reconstruction of a Roman wine press unearthed in Doci near Kaštel Gomilica (Source: Želimir Bašić - Dalmatinska vina kroz stoljeća, Šibenik 2001).

into the vessels used for fermentation. In the early beginnings, wine presses were not used, rather, grapes were placed into sacks and crushed manually. Later on, a press with a wooden lever was employed. One such wine press was unearthed in Doci near Kaštel Gomilica that was at the time a part of a Roman hamlet. It is believed that this press was used for both wine and oil, although it is somewhat different from the Roman presses discovered in other provinces. It consisted of a wooden beam fixed on one end and loose on the other. The loose end could be lowered using a rope wound on a pulley. In this way, pressure was exerted on the grapes laying on the stone slab. The juice flew through a round channel and was gathered in a special stone vessel. The "semi-wine" was produced by diluting grape juice with water, and was often given to the slaves or employees on the estate.

A large ceramic pot called *dolium* capable of storing a volume between 10 and 12 hl was used to preserve young wine, along with a wooden barrel called *cupa*. For a long time, it was believed that wooden barrels were discovered long after the Romans. However, this cannot be true. A picture of a wooden barrel (*vacuae cupae*) from the Roman period was found on a tombstone in the field between Kaštel Sućurac and Solin. Long barrels were most probably tightened in the beginning and at the end with thin elastic birches that were later replaced with iron rings.

Amphoras that could hold volumes between 20 and 26 liters were used for the aging of fine wine. Fermentation of must lasted until the spring equinox, when young wine was decanted into smaller vessels and poured every 18 days. Weaker musts were fortified by adding honey and boiled must. The extra acid was removed using plaster of Paris or calcium carbonate. Purification of wine was performed by filtering through cotton cloths, clarification by adding salt and pigeon eggs. Honey was often added to mask the flaws of wine. Some old procedures in wine production have remained unchanged to this day in small households in the Kaštela region and its hinterland.

The rule of Croatian dukes

Under the influence of disastrous political decisions and the degradation of society in the 4th ct., the cultural and economical decline of the entire Roman Empire began in 476. In their invasions, the Avar and Slavic tribes devastated the Byzantine provinces. At the time, the major city center was Salona; however, conditions in the city were very rough. The situation was best described by the priest and chronicler, Toma Archdeacon:[11]

"Chaos overcame the city...verbally, everyone was brave, but when it came to facing the enemy, they all escaped."

Under such conditions, the Romans were unable to defend the city from barbaric invaders, whose aim was to swiftly overturn the Roman rule, destroy the city's palaces and churches and then burn the rest. The treatment of residents was not much different; those that weren't able to flee to the islands were either captured or killed. The invasions of barbaric tribes were marked by raiding, plundering and looting of rich Roman cities and agricultural estates, including vineyards. It is assumed that the worst fate was brought down on the vineyards along the coast, including those in the Kaštela field and the hinterland. The vineyards on the islands were more or less spared because Romans from the continent sought refuge there. While the mighty and glorious Salona was wiped off the maps, Trogir wasn't significantly upset by barbaric invasions.

During the big migration of peoples, Slavic tribes settled on the eastern shores of the Adriatic. The famous historian Herodotus recorded that the Slavs were already known as excellent farmers, diggers, ploughmen and that they "worked the fields with love".[12] Although they were considered to be mostly farmers and ranchers, the Croats, one of the Slavic tribes, soon adopted the culture of the vine. There are few records from this period related to viticulture, probably because most of the vineyards had been previously destroyed. However, it is certain that the Croats, right after their arrival on the Adriatic coast, completely abandoned their traditional drink, mead, and turned to narcotic wine. Their efforts in expanding the vine culture were very successful, aided by the christening that followed soon after their settling and the use of wine in church ceremonies. It is true that the use of bread and wine is common in many nations and religions, but this custom in Christianity has an especially deep meaning. Jesus Christ ordained that bread and wine represent his body and his blood and should be used during the mass in memory of his name.

After the destruction of Salona, the Croats settled on the fertile land between the ruins of Salona and Trogir in the 7th century. They built their homes on the sloped terrain of mountain Kozjak and formed the medieval communes of Putalj, Kruševnik, Ostrog, Bijaći, etc. In Trogir and other Dalmatian cities, the Latin inhabitants remained after recognizing the rule of the Byzantine emperors, but they were obliged to pay their tributes to the Croats. Trogir had to pay Croatian dukes a yearly tribute of 100 dukats, as well as dues in the form of wine and vine products, according to the agreement signed between Emperor Vasilije Ist and Croatian duke Zdeslav in the 9th century.[13] The separate emphasis on "wine and vine products" clearly points to the importance allocated to winegrowing at the time. In the Kaštela region, the Croats expanded the viticultural areas and raised new vineyards, while the native Roman inhabitants probably continued to work their original fields.

One of the earliest confirmations of a continuing viticultural tradition in Dalmatia after the fall of the Roman Empire was preserved in the testament of citizen Kvirin.[14] To the church of St. Lovro in Trogir he granted not only gold and silver but also vineyards "on the islands and on the solid ground", including those in the Kaštela field. However, according to certain sources[15] (S. Ožanić), the Croats and the other Slavic tribes did not pay enough attention to the aging of wine, and often, its quality was lesser than that allowed by the grapes.

During this period, classical taverns were still nonexistent in Dalmatia. For the most part, wine was kept at the ground level of buildings that often simultaneously served as a storage place for various goods. Because the barrels were not properly sterilized, wine spoilage was a common occurrence.

The founding of the Croatian state was marked by a new revival of the Trogir-Kaštela region, because the residence of state rulers was situated in Bijaći next to the church of St. Marta at the western rim of the Kaštela field. This is where official documents were issued and where the christening of Croatian dukes took place. According to written documents,[16] especially gift certificates, wine was given special attention in the Croatian courts. During the reign of duke Mutimir, the court staff included a cup-bearer, (*pincernarius iupanus*) whose duties included the maintenance of the wine cellar and the chosing and serving of wine inside the court. Since the residence of Croatian rulers was located close to the Kaštela field, it seems logical that the court cup-bearer was in charge of the neighboring vineyards and that Croatian dukes themselves enjoyed Kaštelan wine. A preserved charter[17] from 852 records that the Croatian duke Trpimir also spent time at the residence in Bijaći. As evidenced by various gift certificates and charters (the gift certificate of duke Mutimir from 892, signed in Bijaći in the presence of the cup-bearer Željidjed; the charter of Petar Krešimir from 1066, signed in Nin in the presence of the cup-bearer Djedovit), the court cup bearer often accompanied the duke on his inspections of the estates.

Very early after the formation of the Croatian state, during the reign of dukes, viticulture represented the most significant agricultural activity, as evidenced in the work "Ruling the empire" (*De administrando imperio*)[18] by Konstantin VII Porfirogeneta. In his book, the author mentions that during the reign of Duke Zdeslav, from 878-879, the Roman city of Trogir paid a yearly tribute of 1000 dukats (the tributes of the cities of Split and Zadar were 2000 and 1100 dukats, respectively) in addition to "wine and other valuable goods". The emphasis on wine as a separate tribute reveals the significance of this product in the Trogir-Kaštela region.

Croatian duke Mutimir and his cup-bearer shown in a visit to the estate (Source: Želimir Bašić).

The ownership of land and vineyards raised on the land was often regulated separately. It was not uncommon for the early Croats to rent the land on which they intended to grow vine. In such cases, the land remained in the ownership of the landlord who was given priority in redemption of the estate if the tenant, the vineyard owner, decided to sell it. Situations whereby the land and the vineyard had separate owners were common and not at all unusual and represented the making of an early feudal system. The contracts between the tenant and the landlord were obligatory and often defined in detail the amount of wine or money that the landlord was entitled to. The portion of the yield given to the landlord was called *terraticum*, and it varied drastically in different places.[19] The *terraticum* for the estate Pod Ostrogom near Trogir in 1249 amounted to one fifth, on the island of Šipan near Dubrovnik it was one fourth in 1252 and changed to one third in 1273. The female convent of St. Mary in Zadar was the most benevolent to its tenants, as it required only one tenth of the overall yield obtained in Biban in 1277. The contracts regulating the renting of land were drawn individually or jointly for a certain region. For example, in 1277, the convent of St. Kršovan signed 70 individual contracts.

All sources point to the conclusion that the Croats in the Kaštela field and Dalmatia slowly took over the wine industry and competed with the native Roman inhabitants who prevailed in many Dalmatian cities. The hostility of the Romans towards the tribal newcomers could not stop the integration of the Croats into vital city centers.

Viticulture and winemaking in the free city of Trogir

The medieval city-states, organized as separate economical units, had a significant impact on the development of viticulture in Dalmatia. The main agricultural activities of free cities were viticulture and winemaking.

Wine was very important in medieval Trogir and the prices of must and wine from the Kaštela field and the surroundings were determined by the duke himself. After looking into the wine list composed by his advisers, on a regular basis, during October and November, the duke proclaimed new prices for must and wine (on November 3rd 1627, Duke Alexander Diedo set the prices for must to be 9 liberas, and for a 60-liter barrel of wine 10 liberas).[20]

"The duke threatened those who would not abide by his instructions with the taking away of their must and a sentence to prison." In Split, the prices of wine varied according to the yield, the demand and the offer. Certain contracts regulating the act of sale of wine have been preserved and may be used to reconstruct the role of wine in medieval communities. In one such document[21] dating from the 11th ct., Gribica gave Peter two coins and a barrel of wine in exchange for a part of his house in Solin. The same buyer purchased from Vilko Mira and his brother, Predo, a vineyard beneath the church of St. Maxim in exchange for 12 barrels of wine and a sow.

The extent to which the vine was protected in ancient Trogir may be witnessed from the clauses included in the Statute of the free city of Trogir from 1322 (ar-

The main square in Trogir in the 18th century (Source: Želimir Bašić).

ticles 30, 31, 32)²², whereby damaging of the vines and vineyards was severely punished. In such cases, the city would fine the perpetrator 20 liberas while he had to pay an additional 40 soldas to the owner. Surprisingly, the fine for the damage of olives was significantly smaller, only 5 liberas. If the perpetrator could not pay the fines, he was "permanently defamed and whipped from door to door." If the perpetrator was caught in action, the worst possible punishment was imposed - the cutting off of the right hand.

During the harvest and post-harvest activities, the meetings of the Great Council in Trogir were not held. The court orders were also disregarded, if the matter did not deal with a criminal act. All city council duties were placed aside and all the attention was focused on the vineyards. Medieval cities protected their wine production by special ordinances included in the statutes, forbidding wine imports.

In 1420, the Venetians subjugated Dalmatia and by the middle of the XV. ct., they began to disregard the city ordinances. They endangered the cities' monopoly on wine sales by allowing the import of foreign wine inside the city walls. This is why many Dalmatian cities competed for the privilege to export to Venetian towns. This privilege was first granted to Šibenik in the form of a waiver of 50% of the regular customs and, in 1420, the same rules applied to wine exports from Trogir, Korčula and Hvar.²³ The choice of Dalmatian towns that were allowed to export their wine did not depend merely on the political relations with Venice, but largely on the quality of their products.

The regulations aimed at protecting domestic wine point to the status of this product in medieval Dalmatian communities. Most of the cities abounded with wine, but only had scarce reserves of other agricultural products. This is why medieval Dalmatian towns often experienced economical difficulties in the years of poor harvest. In order to improve the situation, during the reformation of the Statute of Split (article LXIX)[24] it was ordained that new vineyards will not be allowed "beyond the region between Visoka and Sveta Marija of Žnjan." However, "if within this region existed land that was unsuitable for crops, then vineyards could be planted there with the permission of the duke or his substitute. Before granting the permission, expert noblemen have to be sent to inspect the land and confirm, under oath, that it is not suitable for crops." If anyone disregarded the ban, and planted a vineyard where it was not allowed, he was fined with 25 liberas. Severe rationing of land was dictated in the Kaštela field, where it was forbidden to build houses outside of the village communes. "Each inch of land was subordinated to the needs of the locals."

In the medieval period viticulture suffered from the devastation caused by various wars and conflicts, especially those between cities. The disagreements regarding the ownership of fertile land and vineyards located near St. Vital sparked a conflict between Trogir and Split in 1243. When the residents of Split realized that only a fight with Trogir could ensure ownership of the land they felt they were entitled to, they chose the Bosnian duke Ninoslav for their leader. His well-equipped army, along with the formations from Split synchronously invaded, plunged and burned the Kaštela field. A consequence of this war was the devastation of most vineyards in the Trogir-Kaštela region, but not the surrender of the city of Trogir. In the meantime, Butko Julijanov, the duke of the nearby Klis fort, came to the aid of Trogir and the formations from Split had to return empty handed. The war between Trogir and Split lasted for two years and economically and psychologically completely drained both sides. The expenses for weapons and arms, especially ships, could not be paid without the income from wine, so a truce was finally signed in 1244.

Trogir and Kaštela under the rule of Venetians and Turks

After the Venetians subjugated Dalmatia in 1420, Trogir and the Kaštela field succumbed to their force as well. This period was special in the life of Trogir, Kaštela and the whole of Dalmatia. Under Venetian rule (in the mid XVI ct.), Dalmatia numbered about 100,000 inhabitants distributed among 13 cities, 300 villages, 13 towns, 5 forts and several towers. In addition, 12 islands were populated, while 500 villages had been taken by the Turks. In the beginning, Venice left it up to the domestic inhabitants to govern the cities and only appointed its men to the high positions, close to the duke. It is interesting to note one regulation from the city statues, whereby the Venetian duke (Comes) "was not allowed to accept any gifts under no circumstances, except for grapes..."[25]

Under the rule of Venetians, although the planting of vineyards was discouraged in most of Dalmatia, the Trogir-Kaštela area was spared from such protectionist regulations. New vineyards were planted and viticulture, once again, became the dominant agricultural activity in the Split surroundings and the main

Turkish invasion of the neighboring fort of Klis, in 1648. During numerous Turkish attacks, the vineyards in the Kaštela field, shown on the left, were completely devastated (Source: Archeological museum, Split).

source of income to the locals. Also, many noble families began investing in viticulture, because winegrowing became a status symbol. In order to protect the vineyards from looters, the noblemen built small fortresses for protection in the middle of their estates all along the Kaštela bay.

In 1487, two noblemen from Trogir, Jerolim and Nikola Vituri, built a fortified castle with an inner court at the edge of their vineyard properties, close to the sea. People flocked to the castle and gradually expanded the small center to a town that is today Kaštel Lukšić. Another nobleman from Trogir, Pavao Cipiko, in 1512 also built a castle with towers next to his vineyard with the aim of protecting his fruitful vines. Stijepan Stafileo, also from Trogir, built a large house on his property near the coast from which he could oversee the work in his vineyards. His family was noted for their seal, which carried a picture of a vine and grapes in the middle.

In other places at the edge of the Kaštela field, castles and fortresses were built around the same time. The town that is today known as Kaštel Kambelovac was formed by expansion of a large estate owned by the nobleman Cambi in the 15th century. In the 16th ct., a fortress was built on the island called "Gomila", around which a small town gradually developed and was subsequently named Kaštel

Gomilica. All seven Kaštela were built in the style of renaissance summer residences with wide, richly decorated balconies facing the sea and forts with towers and embrasures facing inland.

During Venetian rule, the strategic interests of Dalmatia were often left aside and viticulture was neglected, as were most of the other potentially profitable activities. Wine (just like cloth and salt) was made liable to duty, and regardless of wine surpluses, special export permits had to be obtained. This was the reason why viticulture declined in many regions, although Venice was at the time considered to be one of the most prosperous countries in Europe.

The state of agriculture in Dalmatia during Venetian rule may be best reconstructed from reports compiled by two inspectors, A. Died and B. Giustinian, presented to the Venetian Senate in 1552. For the Trogir-Kaštela region, one report stated:[26]

"Trogir and Kaštela, all the way to Kaštel Kambelovac, produce enough wine for 6 months. They have the quantities necessary for domestic consumption, in some places even more than that, while the oil and fig products are scarce because olives and figs froze in 1510. Besides agriculture, this region is famous for successful cattle breeding."

The records of the Italian scientist and priest, Albert Fortis, who extensively traveled through Dalmatia from 1765 to 1791 and recorded his observations, are also very insightful. He described his journeys in two publications (Viaggio in Dalmacija I-II, Venice 1774), where he stated the situation and circumstances in agriculture. Among his observations, he recorded:[27]

"The plantings of olives and vines are most frequent between Trogir and Split. The seven Kaštela together account for about 13,000 barrels of olive oil and 50,000 barrels of highest quality wine. The Kaštela region is also rich with almond trees. The figs' yield is abundant and amounts to about 300,000 pounds. The rye fields are also scattered around the Kaštela field. The Kaštela hinterland does not produce oil, and wine is produced in scarce amounts. The sheep there give excellent wool and cheese."

From the observations of Dieda, Gustinian and Fortis, it is evident that the vine was most represented in the Kaštela field. The wine was of high quality, and, often, wine surpluses were exported to the neighboring towns. Successful cattle-breeders in the region ensured excellent meat, cheese and milk products for the local population. Although in smaller quantities, other fruit trees and agricultural crops were also represented and allowed the seven towns of Kaštela to be completely independent from their neighbors for most of their living necessities.

In the second half of the XVth century, the Turkish invaders reached the stone walls surrounding Trogir, and the city community became trapped and restricted to the narrow coastal belt. A large number of surrounding villages were devastated and a portion of the population living on the slopes of Mt. Kozjak, fled to safer locations. Many of them flocked to the Kaštela fortresses and, in order to secure them from the Turks, dug up channels all around the stone walls. Seawater filled the channels and represented an additional barrier to the fierce enemy. A movable bridge was lowered when local residents had to cross over. All of a sudden, the noblemen and their labourers were forced to cohabitate and live a life full of fear and uncertainties, the former in the hope of retaining their properties and the latter in the hope of protecting their patrimony.

In the long period of Venetian-Turkish wars, and during the years between major conflicts, the vineyards in the Trogir-Kaštela region were the subject of numerous invasions of Turkish soldiers. There exists no data confirming that the Turks ever conquered Trogir, although they definitely devastated the surrounding fields on several occasions. This is why the inhabitants of coastal villages often deserted their properties, and fled inland. In his travelogue, the Trukish writer Evlija Čelebi (Čelebija) recorded:[28]

"While our soldiers looted and burned the villages, an armful of bullets from our guns fell on the vineyards and houses. We also looted and plunged Trogir and its surroundings, and cut down all the vineyards and orchards in sight. Vineyards surrounded the city Rinica and the fearful citizens often hid there during our raids. Our soldiers stole three large ships, full of wine, cheese and bread. They took the wine barrels to the fortress and hid them there. It was impossible to save the men, women and children who went to the vineyards to help with the harvest."

Because of frequent Turkish invasions, the farmers and shepherds often carried their guns with them to the vineyards. "Sitting in the field and tending to his sheep, the shepherd had his gun ready, and kept it close to him even during the night".[29] The vineyard duties were often neglected or delayed because the locals feared the Turkish attacks.

The Turks represented a threat not only during the wars but in peaceful times as well. The years when the farmers tending to the vines could not perform the harvest were not uncommon. However, when the border with the Turks was moved to only two miles from Split in 1669, the city lost a lot of its territories, most of them with vineyards, and the farmers were forced to tend to the vineyards that fell under Turkish rule. Since the Turkish landlords asked a much smaller reim-

A detail from the Trogir cathedral portraying viticultural and winemaking activities of the locals (Source: Želimir Bašić - Dalmatinska vina kroz stoljeća, Šibenik 2001).

bursement for the use of their land, only one fourth of the yield, certain winegrowers from Split rented the land from the Turks. However, in 1601 the duke of Split and captain Reiner forbade the Venetian citizens to farm the Turkish land and threatened them with a sentence to work on a galley.[30] The labourers from Kaštel Sućurac, who were also Venetian citizens, did not abide by this order and completely neglected the vineyards of their old landlords in the Kaštela field and, instead, turned to the Turks and their land across the border. As a consequence, a conflict sparked between Split and Sućurac, and unrests in the region had to be continually monitored. After the cease-fire, in one of his inspection of the territory, general Civran concluded that the Kaštela field represents a wonderful example of diligently farmed land and properly maintained vineyards.

The silent testimonies of the importance of viticulture in the Trogir-Kaštela region are the motifs used to decorate the gateways, monuments and columns. The most valued work from this period is the portrayal of vine pruning during February and wine decanting during December depicted on the gateway of the Trogir cathedral, which is the work of sculptor Radovan from the 13th century. An artistically shaped stone relief portraying a bird plucking grape berries, dating back to the early Christian era, has been built into a house in Kaštel Sućurac.

Winegrowing in Kaštela in the new era

When the Turks surrendered the fort in Klis in the middle of the 17th century, the country slowly began to recover and build a new life in peace. However, it remained recorded that, after Klis fell, in his last attack on the city of Trogir, the ruler of the Klis sanjak ruined many surrounding vineyards. However, the enactment of peace brought along a steady development in husbandry. The farmers returned to the fields and the locals worked on restoring the houses and planting new vineyards.

Feeling the need to spread agricultural knowledge, in 1788, Antun Radoš De Michirli-Vitturi, together with Ivan Luka Garagnin, Biskup Stratica, Ivan Banović and Count Vracien, founded an academy in Kaštel Lukšić. In contrast to other academies founded across Dalmatia, the Kaštelan academy was oriented to deal specifically with agricultural problems. Among the more important works published in the scope of the Academy that dealt with agriculture in Dalmatia was a book by Luce Chialetich (Čaletić) from 1799, entitled "The letters from Kaštel Lukšić about agriculture in Kaštela." From the Academy's publications, viticultural practices of the period could be easily reconstructed. For example, the pressing of grapes in Dalmatia was in the past performed in many different ways. Different wine presses were used, most of them very primitive. Manual pressing of grapes, using feet, is still common on the islands and in the Dalmatian hinterland. Often, the old Croatian wooden wine press with a lever was also employed. In his book "Viticulture" from 1924, Ivan Rittig provided a description. The wine press consisted of four beams with a lever placed between them. Pressing was performed using a lever with a weight attached on one of its ends. The lever was lowered by winding an iron pulley. As a result, the lowering of the lever exerted pressure on the grapes placed in the vessel beneath.

The friction between societal ranks in Trogir was very pronounced. Especially hard was the relationship between the farmers — the labourers — and the noblemen — the landlords. Only a spark was necessary for the problems between the two classes to escalate and reach a boiling point. The cause of the conflict between the noblemen and their subjects was the right to the ownership of a field called Oprah, on the west side of the city, where both groups wanted to hold their dancing performances. Simple at the surface, the roots of the problem ran much deeper into the economic situation of the two classes. The farmers worked extremely hard in the vineyards and received only a small portion of the yield, while the noblemen prospered on the fruits of their toil. This particular conflict reached the courts in Venice. In 1740, the rebellious farmers from Kaštela were, tied with ropes, brought to Trogir where the conflict was settled at the expense of their lives. For years to come, the rope hung on the wall of the cathedral as a warning to the other rebels.

In 1797, the fall of the Venetian Republic caused numerous unrests in Split, Trogir, Šibenik and Makarska. As a consequence of frustrations accumulated in the past centuries, the farmers initiated a riot in Trogir. They attacked their landlords, and to the horror of the noble class, killed several of them. The unrests were soon stopped by the new Austrian government that took over Dalmatian territory in a peace pact signed in Camopoformium between Napoleon and the Austrian emperor Franz Joseph, on October 17th, 1797.

Under Austrian rule, the first statistical data related to the agricultural production in Dalmatia were collected, as well as the data on wine production. They were compiled by the court adviser Francesko Staffileo, during his stay in Dalmatia between 1797 and 1798, and published in the "Tabella enciclopedica del regno di Dalmazia" (which was presented at the Vienna court in 1798). According to these data, Dalmatia produced about 700,000 hl of wine, of which 158,560 hl was a surplus (404,000 barrels), while the yearly consumption per person amounted to 200 liters (at the time Dalmatia had 256,000 residents). The world-class expert in viticulture at that time, Antonio Dal Piaza, also presented data on wine production, accumulated during his trip to Dalmatia in the 19th century. According to his records, Dalmatia produced between 700,000 and 800,000 hl of wine. If one compares this datum with later sources, the figures seem quite plausible. However, they were not precise enough. According to Dal Piaza's approximations, Trogir and its surroundings produced between 25,000 and 30,000 hl of wine.

In 1806, at the beginning of the French occupation of Dalmatia, significant advancements were made in husbandry. The French administration worked on improving agriculture in Dalmatia and raising the awareness of its importance. Great plans were drawn for more rational cultivation of the vines, olives, and other Mediterranean cultures. Unfortunately, despite the fact that France had one of the most advanced winemaking industries, viticulture in Dalmatia profited the least during this period. After the defeat of Napoleon and his imprisonment on St. Helena island, Austria once again took over the rule in Dalmatia in 1815. The au-

The traditional vineyards with fork-shaped vines and low growth are still common in Dalmatia (photo by Boris Kragić).

tonomy of Dalmatia was very fragile because the court in Vienna was granted all the regional political power in Croatia. Border crossings had to be paid immediately, while the payment of wine taxes was postponed until January. Each farmer had to be inscribed in the land registry, drafted by chancellor Prendini in 1820. The land registry included specifications related to the type of agricultural activity performed. The taxes were imposed based on the data in the registry, and amounted to one tenth of the total income. The feudal system was discontinued in 1848, but the law did little to improve the farmer's unfavorable position.

The traditional way of growing the vine, dating back to the antic period, was preserved and practiced in Dalmatia throughout the 19th century. Forked-shaped vines were often planted on the edges of tree avenues, and vine tendrils wound around the tree trunks, which served as supports. Because the planting of crops was a priority in fertile land, the vine was often moved to sloped terrain and limestone-rich regions where it had to find its way to scarce amounts of soil through layers of rock. This is when unique plantings of vines in the funnel-shaped holes of the rocky terrain surrounding Trogir, Primošten, Rogoznica and the Dalmatian hinterland emerged. Dry-stone walls were used to surround the area where vines were planted. Because of optimal insolation, the grapes were of exceptional quality and the wine extremely valued. In the limestone-rich regions where there wasn't enough soil, it had to be brought from other places and dry-stone walls built as protection against erosion.

The average annual production of wine in Dalmatia until 1898 amounted to approximately 1,177,978 hl; however, certain experts estimate that it even exceeded 1,250,000 hl.[31] The areas under the vineyards were spread across 200,000 acres (80,000 ha) with the average wine production of 6.22 hl per acre (15.56 hl/ha). In the Split surroundings, annual wine production amounted to 600,000 hl and in Trogir it was about 45,000 hl. The production of wine in the district of Trogir, which included a portion of the Kaštela field, continually increased so that in 1858, it amounted to 15,947 hl, and in 1898, it amounted to 45,000 hl.

The wines produced in Dalmatia were of good quality, although certain authors point out quite primitive production techniques. A conclusion followed that Dalmatian farmers were good viticulturists and poor winemakers. In 1864, on his visit to Dalmatia, A. Babo, the head of the School for Viticulture and Enology in Klosteneuburgh wrote in his report: "Against all expectations I discovered that the vine in Dalmatia is cultivated so carefully that it prospers wherever planted."[32] However, when talking about the wine, he noted: "I rarely drank Dalmatian wine that was free of three wine diseases. Almost all of them were sour, tasteless or stenched of goat skin….if the grapes from this region were harvested at the right time, they could give wine comparable to the highest quality French wines." Despite such criticism, in the annual agriculture-forestry show held in Zagreb in 1891, Marin Vuletin from Kaštel Stari was awarded the gold medal and a certificate of excellence for his Crljenac wine, 1886 harvest.

Oidium, peronospera and phylloxera in the vineyards of the Trogir-Kaštela region

In the Middle Ages, when the wars did not represent a threat, viticulture and winemaking were on a steady, progressive course. Soon, wine became the most prized product and represented the major source of income to the population. Viticulture predominated in the coastal region and on the islands. The vine was planted on the sloped hillsides and in the fields, and especially in karst regions rich with limestone that were unsuitable for other cultures. Because of an increase in wine prices, viticulture spread even more and the vine slowly replaced other crops previously grown in Dalmatia. The situation in the Kaštela field mirrored the state of viticulture in the whole of Dalmatia.

The grapes from vineyards planted in the karst were of better quality in comparison to the grapes from fields. The largest quantities of wine were produced in the Split surroundings, about 237,720 hl. In 1829, all of Dalmatia produced 461,471 hl of wine; in 1830, the production fell to 343,046 hl.[33]

Unfortunately, in the 19th century, unbridgeable troubles befell viticulture in Dalmatia. In 1845, an English gardener noticed for the first time that a disease had attacked all the green parts of his vines, and caused damage to the grapes. The infection was ascribed to the fungus *Uncinulla necator* and the disease was officially named "Oidium Tuckery" after Tucker who discovered it. In Dalmatia it was simply called "trouble" or "lug". After appearing in England, it spread to the vineyards surrounding Paris in 1848, and by 1850 it had already reached Italy and Tiroli.

The troubles of England and France meant good fortune for Dalmatia. While oidium devastated the vineyards in Italy, substantial interest arose in the wines produced in Dalmatia. From 1850 and for the next twenty years, ships loaded with wine barrels sailed off from Kaštelan ports to Venice. The first prosperous period did not last very long because oidium appeared in Dalmatian vineyards in 1852 and moved in the direction from west – east. However, the winegrowers whose vineyards were not immediately infected with oidium, especially those in Trogir and Kaštela, sold their wines at a price four or five times greater than before. Dalmatian wine was sold to the Venetians for 30 fiorines per hl., a very good price considering the fact that in Dalmatia the same wine could only achieve a price of 3-4 fiorines per hl. The successful wine trade encouraged the Kaštelan winegrowers to work on increasing their yields by fertilizing the fields. The sudden boost in exports (small conjuncture) was short-lived because increased exports lasted only until 1857, when the troubles of France and Italy were eased because sulfur was discovered as a good protection against oidium. In the meantime, oidium appeared in Dalmatian vineyards and the recuperated viticulture of Italy suddenly represented a threat to Dalmatian exports. Although the use of sulfur helped in the fight against this fungus, infections did not completely disappear until 1866. Because of great susceptibility to oidium, many autochthonous varieties were no longer planted in the Kaštela field and a very important variety, Marzemina, was lost forever.

In 1854, the production of wine in Dalmatia was significantly smaller in comparison to the figures from 1829 and amounted to 276,720 hl, distributed across the following regions: Zadar – 130,155 hl, Split – 129,589 hl, Dubrovnik – 15,845 hl and Kotor – 1,188 hl.[34]

In Trogir and its surroundings, the production of wine from 1857 until 1860, was the following:

Year	Production in hl
1857	15,812
1859	7,920
1860	7,278

Only a few years after the appearance of oidium, in 1867, phylloxera (*Phylloxera vastatrix*) was discovered in France. Phylloxera was brought to Europe along with American vine imports; it first infected the vineyards in England followed by those in southern France. In the next fifteen years, this pest caused irreparable damage to French viticulture. It completely ruined about 2,500,000 acres (one million hectars) of vineyards and the production of wine dropped from 60-70 million hl to about 25 million hl in 1870. The power of its destruction lies in the fact that phylloxera attacks both parts of the vine, the root beneath and the trunk and leaves above the ground. The devastating effects of phylloxera were soon felt in French winemaking and in order to meet its wine demands, France had to resort to imports. Early on, French merchants realized that Dalmatian wines represented a suitable replacement for the renowned French wines.

Wine export from the port of Split in 1925. France was the destination for most of the barrels, where the French winemakers used the strong Dalmatian reds as foundations for their expensive final products.
(Source: Želimir Bašić - Dalmatinska vina kroz stoljeća, Šibenik 2001).

After the critical period in Dalmatian viticulture caused by oidium had passed (from 1857 until 1867) the vineyards in Dalmatia, and especially in the Trogir-Kaštela region, recuperated fairly well owing to the use of sulfur. Because of the increasing demand from France, winemaking along the Dalmatian coast was given a sudden boost right after the few initial successful trade deals were closed. France was in a delicate situation because it had to satisfy the demands of its own wine market while at the same time fight to keep its external markets, most of which were centered around the French colonies. Expert winemakers from France soon discovered that a pleasant mixture reminding them of the red Bordeaux wines might be obtained by mixing the strong reds from Dalmatia with the lighter French red versions. The use of Dalmatian red wines as foundations became common practice in France. The extent of wine export to France is best illustrated by a quote from the professional journal "Bolletino agrario" (no. 20), dated October 16th, 1882:[35]

"The ships (both French and German) flooded Dalmatian ports. The price of wine these days varies around 12 fiorines per hectoliter. In Starigrad, the highest price was 8 fiorines per 60-liter barrel, until, all of a sudden, so many ships sailed into the port that the price increased to 14 fiorines. In Split, one hectoliter costs 18 fiorines, in Šibenik 10 fiorines, and in Kaštela 17-20 fiorines."

The French demand for red wines adversely affected the development of Dalmatian viticulture, because the planting of white wine varieties and the production of white varietals were completely neglected. The rising tide of viticultural enthusiasm almost drowned the Dalmatian field workers. The First Dalmatian

Enological Society, founded in 1871 in Split, played an important role in the export of wine. Four years later, the Society of Enologists was founded in Trogir, along with seven other wine societies throughout Dalmatia. Unfortunately, the society in Trogir existed and functioned only in the formal sense and its activities went by largely unnoticed.

In addition to France, the export of Kaštelan wine spread to Germany, Belgium, and Switzerland because the European producers did not have any surpluses for export to other countries. During the period of large wine exports, Dalmatia and the entire Kaštela region prospered. The prices of wine were quite high and the winemakers could accumulate substantial incomes; the trade and services (restaurants, wine shops, taverns) related to wine suddenly emerged. The ports revived by hiring a large number of dock workers; barrel production was up and the rent of storage space became expensive. Substantial economic advancements were noticed especially in Trogir and Kaštela, where merchants invested in building luxurious houses close to the sea. Wine produced in the Dalmatian hinterland was often included in exports that could not completely be met by the wine produced along the coast.

The period of large wine exports lasted 20 years and was termed the "large conjuncture". In this time frame, the areas under the vineyards increased and the production of red wines soared. The economical status of farmers improved substantially through direct profit, and, indirectly, through the founding of companies devoted to wine production and marketing. The contact between Dalmatia and other wine producing countries had a beneficial effect on the winemaking in this region. Dalmatian winemakers learned from their advanced French colleagues

The export of wine from the port of Kaštel Stari in the late 1800s (Source: Želimir Bašić).

to work the fields in a more rational manner and process the grapes more efficiently.

Drunk from success, the Dalmatian winegrowers did not pay attention to the dangers lurking from their agricultural practices that, through the past years, became completely focused on the vine. Soon, it became evident that the risks of a single-culture economy were not properly weighed. Because the price of wine significantly increased, while that of rye dropped, new vineyards were planted at the expense of other cultures vital to the Dalmatian economy. Olive trees and pine forests were uprooted and every inch of free land was turned to vineyards. In the Kaštela field, the situation was very much the same. The local field workers neglected their previous activities and turned to the more profitable vine that ensured a safe income. The money earned in the wine business was again invested back into the vineyards. The wine production figures for Dalmatia and the islands, achieved during the period of the big conjuncture were never again repeated and neither was the economical boom. For the period between 1875 and 1891, wine statistics were as follows:[36]

Year	Production in hl	Year	Production in hl
1875	1,212,254	1883	1,295,000
1876	1,389,823	1884	1,148,000
1877	566,754	1885	985,270
1878	1,710,800	1886	972,000
1879	907,960	1887	1,585,000
1880	937,800	1888	1,743,584
1881	980,400	1890	937,000
1882	1,420,250	1891	1,150,000

There are no exact data about the total wine exports to the markets outside of Dalmatia; however, estimates may be made according to available sources.[37] Thus, the annual export to France amounted to between 600,000 and 650,000 hl. In extremely good years, it even reached 700,000 hl, which translated to 15,000,000 forints. An additional 100,000 – 150,000 hl was exported to other markets, including those of the Austro-Hungarian Empire. The participation of Kaštelan wine and wine from the surrounding Trogir regions in the total wine exports was significant. The exports to France, England and Italy did not disturb the sales on the already established domestic markets and markets of the Austro-Hungarian Empire, especially Vienna. It is a well-known fact that the imperial court in Schönbrunn was also supplied with wine from the Kaštela field.

An end to the exports to France did not come abruptly as claimed by several authors, including S. Ožanić.[38] Rather, the steady recuperation of French vineyards slowly decreased the need for foreign imports. A drop in wine exports from Dalmatia was first felt in the mid- 1880s, and it gradually decreased, until 1892, when it completely ceased because France introduced much higher taxes in order to protect the domestic wine production. However, the export of Dalmatian wine to Germany, Belgium, Hungary and Switzerland continued and certain quantities, though not very significant, were also exported to Great Britain and some of its colonies. The export to other European countries was possible because viticulture was not restored there and French winemakers did not have enough surpluses to meet everyone's needs.

The export of Dalmatian wines completely stopped after the restoration of French winemaking, and exportation troubles were magnified by the appearance of phylloxera in Dalmatian vineyards. Mate Dudan, a traveling teacher, first noted the pest on the islands of Olib and Silba. When phylloxera first appeared in 1894, Dalmatia had 194,010 acres (77,604 hectars) under the vineyards. The production per acre amounted to 6.54 hl with the overall production yielding 1,383,201 hl.[39] The vineyards were most concentrated in the regions of Split, Trogir, Brač and Hvar. The district of Trogir, including the Kaštela field, had 8,738 acres (3,495 hectars) of land under the vineyards. The production per acre was 7.72 hl and the overall production, 67,412 hl. Two years later, in 1896, the production per acre dropped to 7.04 hl and the overall production was 61,500 hl, of which white wine accounted for 1,880 hl and red wine accounted for 59,620 hl. An excellent harvest was documented in Dalmatia in 1893, prior to the appearance of phylloxera (1,557,930 hl). In contrast, after phylloxera's rampage in the vineyards, the harvest in 1897 yielded 51% less wine (770,134 hl).

Phylloxera spread continually throughout Dalmatian vineyards, leaving almost nothing behind after its fatal visits. In 1898, it infected the vineyards in the districts of Benkovac and Šibenik. All attempts to veer this pest from its devastating course were unfruitful. The leading Dalmatian expert on phylloxera, S. Ožanić, commented on the situation:[40]

"Phylloxera has taken full swing ahead. Neither fences nor distances can stop it. It overcomes every obstacle and ruins the vines on contact."

By the end of 1900, phylloxera spread across 34,940 acres (13,976 hectars). In 1902, it infected the vineyards in Knin and in 1904, the entire Zadar region. Continuing on its course, in 1905, it attacked the vineyards in the district of Split and the same year its presence was noted in the surroundings of Trogir and in the Kaštela field. From there, it spread across southern Dalmatia and the islands.

Phylloxera completely paralyzed the Dalmatian economy. Most families that were making their living from vine products were brought to the brink of survival. Dalmatian viticulture was in the past troubled by numerous misfortunes that often ended in large economic crises; however, nothing could measure up to the devastation caused by phylloxera. The authorities implemented various measures in their attempt to deal with the situation. First and foremost, they worked on stopping the spread of the infection. In 1895, the first regulations were enacted under the title "The law about anti-phylloxera activities in Dalmatia". The second measure was the enactment of an international treaty signed in Vienna on November 2nd, 1888, designed to regulate the trade of plant material.

Most of the measures came into effect when it was already too late. Many vineyards in the Kaštela field succumbed to the pest, although, for years to come, the vine remained economically the most significant agricultural crop in the region. On the long course that oidium and phylloxera took through Dalmatian vineyards, another disease, the peronospera, also began its journey. It was first noticed by Pavao Zanki in the vineyards surrounding Zadar and later on in Zaton near Šibenik and Ston on the Pelješac peninsula. From there it spread across all the winegrowing regions of Dalmatia. While the merchants and landlords found a way to recover from their troubles, the field workers in Kaštela, once again, found themselves double-crossed. With nothing left to lose, they ventured off into the distant lands of the American continent.

The wine clause and problems of the local winegrowers in the 19th century

During the fierce attacks of phylloxera on Dalmatian vineyards, from 1894-1918, the infamous Wine Clause (1893-1903) additionally complicated the position of the Dalmatian winegrowers. The Wine Clause was an integral part of the trade deal signed between Italy and the Austro-Hungarian Empire on December 7th, 1887, which was further confirmed and prolonged until December 31st, 1903, by the Ship Trade Treaty signed on December 6th, 1891 between the two governments.[41] Special regulations related to wine exports were included in this international treaty, ratified by the "Vienna Imperial Council". Thus, for the import of Italian wines from Lombardy and Venice, special lower taxes were imposed (only 3.20 fiorines per hl) in comparison to taxes for wines coming from other provinces. These regulations together comprised the Wine Clause, which, many argue, Austria-Hungary signed under the political pressure exerted by its European neighbors. The Wine Clause had a devastating effect on Dalmatian wine trade, because Italy reduced its exports to France and turned its attention to other countries, especially Dalmatia in the Austro-Hungarian Empire. From September 1892 until September 1893, Italy exported 1,258,034 hl of wine to the markets across Austria-Hungary – a quantity comparable to the annual wine production in Dalmatia that year. The harvest in 1893 was especially successful with 1,557,930 hl of wine produced. As a result, the taverns in Kaštela and Trogir were packed with unsold wine.

The Wine Clause especially affected the inhabitants of Kaštela because they were most dependent on the vine. The inability to sell wine at an appreciable price caught them by surprise and threw the economy completely off balance. As a consequence, everything else stalled because no incomes from the winemaking industry could be invested into other economical activities. The prices of domestic wines dropped on three occasions because of competition from cheap Italian imports. With the adoption of the Wine Clause, the economic status of Dalmatian farmers suddenly got worse because the export of wines to Germany significantly decreased and the exports to France ceased altogether due to high taxes.

The frequent letters of protest sent from Kaštela and all of Dalmatia to the Imperial court in Vienna adopted a serious tone. The Dalmatian assembly chose seven experts from all parts of Dalmatia to draft a document listing the measures necessary to relieve the hard situation in viticulture. On April 1st, 1892, the assembly compiled a list of 14 steps for the recuperation of Dalmatian viticulture.[42]

Especially interesting were the suggestions aimed at lowering the taxes for wine exports to other countries, the opening of special sea and railroad routes for the transport and trade of wine, as well as the planting of tobacco in places where vineyards were taking too much space. However, these demands were completely disregarded by the Imperial Council in Vienna.

The problem of unsold wine and low prices was stressed in the report of the Wine Society from Split whose members, after the harvest in 1898, turned to Dalmatian counties and demanded statistical data about the yield, prices and the state in viticulture. For the Trogir region, the report cited:[43]

"In Kaštela, the wine was excellent and almost one third of it was sold at a price of 13-14.5 fiorines per hectoliter; however, the present state guarantees no buyers in the future."

With the termination of the international treaty between Austria-Hungary and Italy, most of the measures from this document remained active, except for the Wine Clause. On the day when the termination of the treaty was verified, October 15th, 1904, the ominous Wine Clause died. Once again, Dalmatia was given its rightful chance to compete with its neighbors on the wine market. However, the past several decades of protectionist taxes imposed on Italian wines left tragic consequences on its economy. Dalmatian farmers would never fully recover from the devastation caused by the Wine Clause, phylloxera and oidium. Numerous abandoned vineyards scattered throughout Dalmatia today are silent witnesses of a troubled past.

Vineyard restoration and the establishment of nurseries in Kaštela

Since 1897, when the production of wine in Dalmatia reached the lowest levels ever, the figures gradually began to rise. From 1892 until 1900, the annual wine production in Dalmatia was as follows:[44]

Year	Production in hl	Yield in hl/ha
1892	1,237,530	/
1893	1,557,930	20
1894	1,383,320	17.8
1895	1,126,750	14.5
1896	1,354,980	17.4
1897	770,134	/
1898	922,176	9.9
1900	1,206,494	15.5

From 1900 onwards, wine production more or less averaged 1,000,000 hl per year. In the beginning, Dalmatian winegrowers did not have the knowledge or the means necessary to recover from the damage caused by phylloxera. They observed their neighbors, Italy and France, and gradually learned how to handle vineyard restoration. All new vines were planted on imported American rootstocks, resis-

One of the first wine cooperatives was established in Komiža on the island of Vis. The cooperatives played a key role in vineyard restoration in the early 20th century (Source: Želimir Bašić).

tant to phylloxera. Nurseries were established and schools of viticulture and enology founded within existing faculties. Experts in viticulture were sought after to hold training courses for field workers. New, modern methods of growth were slowly implemented and new cultivars introduced to Dalmatia. The vineyards in karst regions and unapproachable locations were not restored. Rather, new vines were planted mostly in the fields because American riparia rootstocks demanded richer and deeper soil.

In the beginning of the fight against phylloxera, varieties imported from America were own-rooted; however, it was soon discovered that they are not suitable for the climatic conditions of Dalmatia and the wines produced were of poor quality. Luckily, only about 7.5 –12.5 acres (3-5 hectars) of vineyards were lost to such unsuccessful attempts, and in Kaštela, the damage was negligible. The first nursery for American rootstocks was established in 1884 on the island of Silba, where phylloxera was noticed for the first time and where it left behind only miles of deserted pastures. Additional nurseries in the surrounding regions appeared one after the other. The one in Kaštela was among the first of its kind, spreading across 13.13 acres (5.25 ha) of land. The vines in this collection were used in studies of rootstock fertility and the adaptability of native vines to new American rootstocks. Along with the state-financed collections, local winegrowers established their own private ones that often had a special meaning to the family. Most of them were maintained until direct planting of canes into vineyards became common practice. The restoration of vineyards in the Kaštela field and the rest of Dalmatia was performed within a very short time period, much faster than in any other region of the dual monarchy. The customary restoration procedure consisted of the planting of wild vines in the first year, followed by grafting in the second, and harvest in the fourth.

With the aim of prompting mutual cooperation among the winegrowers, the "registers of common wealth" were established as sorts of savings-banks that later turned into the first consumer-producer cooperatives. The first such cooperative in the whole of Croatia was founded on December 1st, 1864, in Korčula (*Cassa di mutuo credito in Curzola*).[45] Soon after, founded according to the regulations defined in the Law of associations, similar cooperatives appeared in other towns. In Kaštel Stari, a cooperative was founded in 1903, followed by the agricultural cooperative founded in Trogir in 1907. Most of the members in such associations were winegrowers, because winegrowing continually represented the most important branch of agriculture in Kaštela. When the Founding Senate of Dalmatian Cooperatives met in Split on February 21st, 1907, fifty delegates assembled from the entire region: Trogir, Kaštel Lukšić, Kaštel Kambelovac and Solin.

The cooperatives aimed their efforts at restoring the vineyards ruined by phylloxera. In the district of Split, 15,500 acres (6,200 ha) were replanted, including the vineyards in the Trogir-Kaštela region. While vineyards in the Kaštela field were being restored, in 1911, phylloxera finally reached those on the island of Korčula. Due to their geographic isolation and relatively scarce contacts with the continent, the vineyards on remote islands remained spared for a very long time.

The assortment of grape varieties in the Kaštela field in the past

The assortment of cultivars in the Kaštela field significantly changed in the past. In general, there exist no significant data about vine varieties cultivated in the early viticultural periods because documenting variety names has become a relatively recent practice. Since the first mention of variety names in the winegrowing countries, including Dalmatia, many cultivars have been completely discontinued from use and only a few endured the passing of time.

Before the appearance of phylloxera, the Kaštela field abounded with grape varieties. Throughout the centuries, the Dalmatian seamen brought home from their journeys many cultivars, which they collected based on their promising attributes of high yield, good quality or ripening period. Adding to this wealth were the autochthonous varieties believed to be native to the Kaštela field. According to S.

Ožanić,[46] due to their documented extensive exploitation throughout the past two centuries, the cultivars Plavina, Plavac mali and Lasina most probably originated in southern Dalmatia. Their long tradition of cultivation in the Kaštela field also points to the Kaštela as their possible original home. It is also indicative that these varieties remained undetected in foreign ampelographies published in the past century.

Many varieties have the additional adjective "kaštelanski" in their name pointing to their origin from Kaštela, or at least a long viticultural tradition in this region. For certain varieties, it is impossible to deduce whether they are of domestic or foreign descent. During the planting of vineyards in Kaštela, little attention was paid to the selection of plant material and it was common for one vineyard to contain many different grape varieties. Ožanić even noticed vineyards with 50 different cultivars in Kaštela. In his publication "Dalmatian ampelography",[47] S. Bulić recorded that about 200 different grape varieties exist in Dalmatia— 90 white, 78 red, and 4 rosé (in addition to 28 described by I.K. Novak), with about 1300 synonyms. In this publication of enormous value, compiled by Marcel Jelaksa and published after Bulić's death, the included ampelographic descriptions have been unmatched even to this day in terms of their scope and accuracy. Born in Vranjic, a village near Split, in 1865, Bulić had his first contacts with the vine in Solin and the Kaštela field. For the purposes of his research, he planted an experimental vineyard in Split containing more than 200 grape varieties collected throughout Dalmatia. In his monumental work, Bulić solved many discrepancies in cultivar identification that had troubled his colleagues in the past, and classified varieties according to the most significant viticultural properties such as yield, ripening period, grape quality, etc.

In 1925, Bulić recorded the presence of a large number of grape varieties in the Kaštela-Trogir region. He noticed the following white cultivars:

Cibib in Trogir
Cibib muškatni in Trogir and Kaštel Novi
Karmelitanka in Kaštel Novi and Kaštel Lukšić
Krivalja in Kaštel Sućurac and Kaštel Novi
Maraština in Kaštel Novi and the Trogir surroundings (Seget, Sitno and Marina)
Muškat mali bijeli in Kaštel Lukšić, Trogir and Sućurac, also known as Muškatin in Kaštel Novi
Pagadebit in Kaštel Novi, Kaštel Sućurac and Kaštel Gomilica
Vugava in Kaštel Sućurac, Kaštel Novi and Trogir surroundings
Žilavka in Kaštel Lukšić

Bulić also recorded the presence of a large number of red grapes in southern Dalmatia, many of which had ten or more synonyms in use. It was later discovered that the names Plavac and Crljenak are in some places used interchangeably, which had added to the confusion in the past. Bulić found the following red grapes in Trogir and the Kaštela field:

Babić in Kaštel Novi and Trogir

Brzamin in Kaštel Novi and Kaštel Sućurac, where it is also known under the name Tribidrag and Brzamin

Dobričić, also known as Crljenak slatinski in Kaštel Novi and Slanac or Šiljaka in Trogir

Lasin in Trogir, also known under the synonym Pošipaj and Pošipaj veliki, Slast and Zlarinština in the Trogir hinterland

Ljutac in Kaštel Lukšić, also known as Bedalovac, Plavac Bedalovac and Ljutun. In Kaštel Sućurac its synonym is Plavac Veliki; in Kaštel Novi it is called Plavac veliki and Ljutun, and in Trogir, Plavac Bedalovac

Ninčuša in Kaštel Lukšić, Kaštel Novi and Trogir surroundings where its local name is Vinčuša

Okatac in Kaštel Lukšić, Kaštel Novi, Kaštel Sućurac, Trogir and its surroundings. In these places, the locals also use names Glavinuša, Brgljun and Pošip crni

Plavac mali in Kaštel Sućurac and Kaštel Novi, where it also called Crljenak; Kaštel Lukšić, where it is known as Crljenak, Crljenak pucavac and Crljenak smijavac. It was also present in Trogir and the surroundings where it went by the names Crljenak, Crljenak trogirski, Crljenar and Crnjak

Plavac mali kaštelanski in Kaštel Sućurac, Kaštel Novi and Trogir, where it was known as Rudica

Plavina crna in Trogir and its surroundings (Trolokve) and in in Kaštel Sućurac, where it is known as Modrulj and Plavac

Pljuskavac in Kaštel Novi, Kaštel Lukšić and Trogir and its surroundings (Sitno, Sratok, Suhidol and Marina), where it is also called Rogoznička Šiljaka, Rogoznička velika and Rogoznička mala

Soić crni (Plavac veliki crni) in Kaštel Sućurac, where it also goes by the name Bračanin, and in Kaštel Lukšić, where it is called Glavinuša Bračka

Rumanija rumena in Kaštel Novi and Kaštel Sućurac, locally known as Rumenka.

Although the planting of vineyards in Dalmatia was never planned, and although the winegrowers often did not follow the experts' suggestions, they often performed selections of their own. The largest selections in Dalmatian vineyards and vineyards of the Kaštela region were performed on four occasions.[48] The first selection began around 1860, with the appearance of oidium in Dalmatia. Many varieties susceptible to this fungus were excluded from further planting and replaced by more resistant ones. The second selection occurred between 1875 and 1890, when exports, especially those to France, increased. This was when the red varieties yielding deep colored wines similar to those of the French Bordeaux region became dominant in the vineyards of Dalmatia. The third occurred after the appearance of phylloxera, when the natural selection process caused the disappearance of many traditional cultivars. The last massive selection process took place during vineyard restoration, whereby new vines were mostly grafted onto resistant American rootstocks. In contrast to the previous three selections, this last one was performed systematically and supervised by the experts. Due to funda-

Red wine grape varieties still prevail over white cultivars in the Kaštela field (photo by Boris Kragić).

mental changes to the assortment from the mid 1800s until today, many old varieties have been either completely lost or reduced to symbolic numbers.

The attempts at introducing certain European varieties such as Pinot noir, Kadarka and Rhine riesling to vineyards in Dalmatia and the Kaštela field have not been successful. These varieties were soon abandoned because native cultivars gave higher yields and better quality grapes. Only the variety Žilavka from neighboring Hercegovina adapted well to the climatic conditions of the Kaštela field, but even this variety has in the past several decades also been abandoned in favor of certain Dalmatian varieties. Those varieties that were incompatible with American rootstocks were eliminated from the start. Also, the viticultural experts insisted on monovarietal vineyards and the return to the native varieties, grafted onto American rootstocks. Some varieties were discontinued, not only because of their susceptibility to diseases and pests, but also because they could not guarantee an income due to poor properties.

The number of wine grape varieties grown in the Kaštela field was significantly reduced after phylloxera. Only about a dozen cultivars that had previously proved their qualities were kept and planted across significant areas; namely, Babić, Plavac mali, Maraština, Ninčuša, Okatac, Kuč and a few types of Crljenak. In the post-phylloxera period, several wine cellars were already established in the Kaštela field. The most famous one, praised for its high quality wine, was the property of the Šimeta family in Kaštel Stari. This fairly large private winery had the capacity of 100-120 wagons (1 wagon ≈ 22,000 pounds) and the wine was largely stored in oak barrels. Some of the barrels have been preserved and the date engraved on

111

*Wine fair in the neighboring town of Solin in the late 19th century
(Source: Želimir Bašić - Dalmatinska vina kroz stoljeća, Šibenik 2001).*

them points to the year of production, 1886. It is assumed that the most famous Šimeta wine cellar was built around that time as well.

The absence of wine surpluses in the Kaštela field

After the end of the First World War and the disestablishment of the Austro-Hungarian Empire in 1918, Dalmatian wines lost their traditional markets. For their placement at strong consumer centers (Vienna, Prague, Graz, etc.), complicated administrative steps had to be taken and high taxes paid. The taxes posed special difficulties to the winemakers from Kaštela who already had traditional buyers in Vienna. The new government centered in Belgrade did nothing to aid the wine exports; on the contrary, it often caused further delays by implementing unreasonable measures, such as additional export taxes that subsequently made Dalmatian and Kaštelan wines too expensive. The winegrowers were once again faced with a crisis because Dalmatia mostly produced red wines intended for mixing with foreign wines that could not be sold on the domestic market. Many red grape varieties grown in the Kaštela field were planted specifically for export purposes. The Dalmatian winegrowers soon felt the harsh consequences of the govrnemnt's wrongfully oriented viticultural policy. On the one side, the produced quantities of drinkable table wines were insufficient, while at the same, time surpluses of red wines were accumulating in Kaštelan wine cellars.

Although significant efforts were invested in the restoration of viticulture in Dalmatia, the areas under the vineyards never again reached the pre-phylloxera numbers. There were several reasons for this situation, starting with a reduction in the number of winegrowers due to economic emigration of the local population and ending with an unsuccessful introduction of American rootstocks in certain karst-rich regions of Dalmatia. During the wine crises, the number of people who emigrated from the Trogir-Kaštela region was not even close to the thousands that left the islands; however, a change in their orientation from the vine to other agricultural crops and activities, such as olives and fishery, was evident. From this period onwards, the viticultural areas in the Kaštela field continually decreased, and in the next several decades brought the vine to the brink of its existence. In 1922, only 84,803 acres (33,921 ha) of vineyards were renewed, which amounted to about 26% of the total viticultural areas.[49] In the district of Split, which administratively included the Kaštela field, out of the 34,595 acres (13,838 ha) ruined by phylloxera, about 54% or 18,750 acres (7,500 ha) were restored. In their attempts at renewing viticulture, the local winegrowers were largely left to themselves. Often, their incentives were counteracted by unfavorable government measures. In their efforts to sell the wine, on numerous occasions, the merchants exported it at a price much below its real value. The government stimulations for exports were completely useless to the winemakers because high taxes in the end left them without income. According to official statistical data presented by S. Ožanić,[50] in 1938, 844,000 hl of wine were produced in Dalmatia, a much smaller quantity than before the appearance of phylloxera. In the district of Split that included all the vineyards from the Trogir-Kaštela region, the production dropped from 255,000 to 123,536 hl. Vineyards were spread across 129,250 acres (51,700 ha) with the average yield of 17.94 hl / ha. As far as wine consumption is concerned, at the time, 400,000 hl of wine was consumed in Dalmatia, which amounted to 70-80 liters per person (3.5 hl per family); about 200,000 hl was consumed in the taverns and about 320,000 hl was exported.

Up until the Second World War, agriculture still represented the main activity in the Kaštela field. However, after 1945, as a consequence of wartime destruction, lack of manpower in the fields, and disrupted farming activities, the vineyards suffered the most. As a result, the areas under the vineyards were reduced by an additional 40%, or 33,750 acres (13,500 ha), with respect to the pre-war period. In the preserved vineyards, fertility severely decreased and in 1945 only about 400,000 hl of wine was produced, which corresponds to about one half of the pre-war production.

The period between 1940 and 1945 was in the first place characterized by a lack of field workers because many young men participated in the war or emigrated from their hometowns. Also, the vineyards were ruined as a consequence of wartime actions and subsequent plundering. It was almost impossible to organize a stable wine market, and speculations with the rare wine surpluses were common. Socio-political and economic relationships changed after the Second World War. Collectivization and the formation of farmers' cooperatives, very much similar to those in the Soviet Union, were underway in Dalmatia. However, mechani-

Harvest in Dalmatia at the turn of the 18th century
(Source: Želimir Bašić - Dalmatinska vina kroz stoljeća, Šibenik 2001).

zation could not be implemented in most Dalmatian vineyards because they were too small and the terrain was unapproachable. Forced industrialization in the cities caused massive migrations of people from villages and towns, especially those on the islands, to continental cities. As a consequence, the seven Kaštela were additionally robbed of their manpower.

With the aim of better placement of Croatian wine on foreign markets, on November 10th, 1946, the state company Vinalko was founded.[51] The wineries spared from war devastation, including those in Kaštel Sućurac, were merged with Vinalko and counted approximately 500 wagons of storage space. Soon after, the Split branch of Vinalko separated and acted as an independent company with wine cellars in Kaštel Stari, Kaštel Novi, Komolac-Dubrovnik, Grude, Šibenik and Ćilipi. Owing to Vinalko, for the first time after the war, wine was again bottled with labels designating geographical origin. The bottling was performed largely manually since mechanization of the process was too expensive. In the post-war period, the average production of wine in Dalmatia was as follows:[52]

Time period	Production in hl
1944-1946	436,666
1947-1951	562,920
1952-1956	749,494
1957-1958	933,499

Respecting the winegrowing tradition, in 1943, the state authorities reopened the Agricultural high school located in the Vitturi castle in Kaštel Lukšić with specialized courses for education in viticulture. Although many students graduated from this high school, only a few of the experts remained in Kaštela.

In 1964, 92,500 acres (37,000 ha) were under the vineyards in Dalmatia, which amounted to about 19% of all agricultural areas, yielding about 1,000,000 hl of wine. The wine quantities produced in Dalmatia corresponded to about 40% of the total wine production in Croatia. According to the performed analyses, each fourth resident of Dalmatia made a living from the vine products. The consumption of wine in Dalmatia was high and amounted to approximately 50 liters per person.

In the 1960s and 1970s, what remained of the small private vineyards in the Kaštela field was replaced by industrial complexes (cement factories, ironworks) and, later on, an airport. While modern wineries and large plantations were raised in certain places in Dalmatia, no new investments in winemaking were made in the Kaštela surroundings. The existing wine cellars were slowly converted to serve other purposes. Up until 1980, sporadic wine surpluses in certain regions reached up to 13 wagons of grapes per day, only to completely disappear in the 1990s.

Today, vineyards in the Trogir-Kaštela region are rare sights. According to recent statistics, 192,500 acres (77,000 ha) of cultivable land exists in this region, but vines are planted across only about 2,913 acres (1,165 ha).[53] All vineyards in the region today are private property. There are no significant wine surpluses and the wine cellar in Kaštel Novi run by Dalmacijavino remains empty for the most part. The grapes obtained from the Kaštela vineyards are mostly used for wines produced for own consumption. Some farmers from Kaštela also buy grapes from other regions and produce small quantities of wine in their own wine cellars.

The Kaštela field has lost the image of the green carpet it used to have during viticultural prosperity. The traditional bond between the field workers and the vine has also disappeared. The rare vineyards situated between apartment buildings and industrial sites are silent witnesses of passed centuries when the vine ruled the region. Only time will show whether the new generation will, at least partially, restore the old glory to the Kaštela field.

CHAPTER FOUR

A SPECIAL KIND OF A WINE STORY

Based on the present state of viticultural activity in Croatia (both the acreage under vineyards > 13,000, and the assortment of wines), central and south Dalmatia, with its islands, are definitely the most significant.[1] While there is a prevalence of introduced varieties in the continental region (cca 50% Italian Riesling, cca 15% Chardonnay, cca. 7% White Riesling), autochthonous varieties dominate in Dalmatia and on the islands (mostly Plavac mali, Babić, Vugava, Pošip and Debit). Among the more than 100 autochthonous grape varieties present in Croatia today that were mentioned, about 60 are found in the coastal belt.

Today, several cultivars with a long tradition in Dalmatia stand out from the rest. The wines made from the grapes of Plavac mali, Pošip bijeli, Grk bijeli, Maraština, Vugava bijela, Malvasia dubrovačka, and Babić are unique, each one a story in itself. The Babić vineyards in Primošten, laid upon layers of rock and surrounded by kilometers of dry-stone walls, are a tribute to the vines' struggle with the land and, probably, the most impressive wine story that Dalmatia has to offer. An appropriate ending to a book that aims at turning the reader's attention to the viticultural treasures of Dalmatia, among which Zinfandel has recently found its place high on the priority list, also warrants a closer look into the "saga of Primošten vineyards".

❧❧❧

Pošip bijeli - the high quality white wine cultivar of Korčula (photo by Boris Kragić).

Pošip bijeli is the predominant white wine cultivar in Lumbarda on the island of Korčula (photo by Boris Kragić).

It seems that the vine has been associated with the coastal region of Šibenik and Primošten for as long as one can remember. It is well-known that the Illyrian tribe, the Liburni, cultivated the vine in this region in the 8th ct. B.C.[2] The tradition continued with the Greeks and Romans up to the great migration of peoples when the Croats settled in the region in the 7th century A.D. They slowly turned to the vine for prosperity, learned to live with it from season to season, and treasure its loyalty.

Šibenik is a picturesque small town, located at a strategic site on the east Adriatic coast and in the past served as an important port. It was founded after the establishment of Croats on the Adriatic coast, in the 11th century A.D. Today, this small town, with one of the most interesting city centers on the Adriatic, plays an important role in tourism, and is the meeting point for people seeking refuge on one of the beautiful islands in the Šibenik archipelago (Zlarin, Prvić). Although Šibenik relied on aluminum, chemical and textile industries in the past, its inhabitants respected and encouraged the culture of the grapevine from the very beginning. Archeological evidence pointing to a grape-growing civilization in the Šibenik Diocese dates from the 11th ct. A.D. Numerous stone fragments were unearthed in the Šibenik surroundings, decorated with plaits of wine leaves. The first written document mentioning vineyards comes from 1298, when the Great Town Council transferred the ownership of a large number of its vineyards to the Church.[2] The ornaments found today inside the Šibenik Gothic-style cathedral of St. James (Sv. Jakov), built in the 15th ct. (1431-1536) by master Juraj Dalmatinac, feature vine leaves and strings of grape clusters.

Dry-stone walls surrounding the Babić vines in Primošten. Among a myriad of international entires, the Primošten vineyards took the first place and won the gold medal at the World exhibition of gardens, Flora 2000, held in Japan (photo by Boris Kragić).

With the passing of time, the evidence of a grape-growing population in the Šibenik surroundings became more abundant. According to Ž. Bašić,[3] the specific style of dry-stone wall use in the design of Primošten vineyards probably dates back to the 16th and 17th centuries, in the time of the Venetian-Turkish wars. This was a time when the people withdrew from large towns and settled in smaller vil-

Supreme quality Babić wine produced by the Vinoplod winery in Šibenik (Source: Želimir Bašić — Vina Dalmacije, Split 1999).

lages in more remote places and on the islands. There they found fertile soil in scarce amounts and were forced to adapt their winegrowing style to the rockier, steeper grounds. Planting vineyards under such conditions was twice as hard as on some other locations; boulders had to be broken apart mechanically and stones arranged into walls, piece by piece.

The local winemakers claim that the dry-walls serve a dual purpose. Besides preventing erosion of land and adding to the beauty of the landscape, the stones surrounding the vines reflect sunlight off their white surfaces and in periods of drought, store moisture. Upon drawing nutrients deep from within the barren land, the vines easily sustain longer dry-spells, common in Dalmatia. The predominant variety of Primošten, Babić, is best accustomed to this region and is well-known for its high quality red wines. In the past, Babić was often found mixed with other varieties in smaller quantities; however, after restoration of vineyards on resistant American rootstocks, Babić spread across the entire northern and southern Dalmatian winegrowing regions. The Babić vine is characterized by a well-developed hard cane and moderate vigour.[4] Babić grapes reach technological ripeness rather late, in the second half of October.

The viticulture region around Primošten today produces considerable amounts of wine; however, only the highest quality grapes are used for the production of outstanding Primoštenski Babić (Vinoplod winery, Šibenik). This wine typically contains 12-14 vol % alcohol; its dark red color and unique, soft bouquet are achieved by careful aging in oak barrels. It is served with fish specialties, red meat and first class cheeses and is widely distributed in Croatian stores.

Professor Vinko Tadijević calculated that it would take one man twelve centuries to raise the dry-stone walls around the 125 acres (50 ha) of vineyards in Primošten.[5] Thousands of kilometers of dry-walls were actually built in less than one hundred years, from the beginning of the 18th to the beginning of the 19th century. It is unknown how many field workers participated in this grandiose project, which, many argue, borders on fanaticism. One thing is certain, each stone, and there are millions of them, was handled with utmost care and precision – only a view from the air gives the whole picture.

The Babić vineyards in Primošten. Inscribed in the UNESCO World Heritage List, the dry-stone walls of Primošten are an unique tribute to the vine in Dalmatia (photo by Boris Kragić).

This immense human effort on the part of local field workers was recognized in several ways. A picture of Primošten vineyards was placed in the United Nations Building, as a symbol of the beauty of cultivated landscape and the effort at sustainable development. At the world exhibition of gardens, Flora 2000, held in Japan, Croatia presented itself in the Nations Gardens' sector with a symbolically styled Primošten vineyard and the Velebit mountain rock garden. The Croatian garden won the Gold Medal, and the only one among other foreign entries, the Excellent Prize. Following this international recognition, the Committee for the Regulation of Space and Environment Protection of the Croatian Senate, dated 23 May 2002, joined the initiative for inscribing the Primošten vineyards in the UNESCO World Heritage List.[5] The resolution reads: "Full support is being given to all activities related to the inscription of the Primošten vineyards in the UNESCO List, in the Protected Cultural Landscapes category". A team of Croatian-Japanese specialists, sponsored by the Ministry of Culture of the Republic of Croatia, is working on a survey intended to show support for this initiative. The vineyards of Primošten are an extraordinary example of human work and natural beauty, and as such, deserve their place alongside other protected cultural sites inscribed in the UNESCO World Heritage List. In a world that is increasingly distancing itself from nature and interrupting the symbiosis between the wild and the tamed, the Primošten vineyards stand out as a constant reminder and a warning sing.

ಸಿ ಸಿ ಸಿ

The barren soil of Dalmatia, hidden deep below the sun-bathed and wind-blown layers of karst, is where the vine found its home. When grown under the climatic conditions of coastal Croatia, at specific *terroirs* exposed to just the right balance of sunshine and moisture, Dalmatian grapes give the best quality wine with characteristic flavors and aromas of blended Mediterranean herbs. These wines are not to be found anywhere else in the world, once tasted, they will keep bringing you back to the heart of the Mediterranean, to the table of a tired Dalmatian field worker.

NOTES

Chapter 1: Winegrowing in Croatia - the legacy of our forefathers

1. Zohary, D., Hopf M. (1993) Domestication of Plants in the Old World, Clarendon Press, Oxford.
2. Amerine M. A., Singleton V. L. (1977) Wine: An introduction, University of California Press, California.
3. Olmo H. P. (1976) Grapes. In: Simmonds NW (ed) Evolution of crop plants. Longman, London p. 294-298.
4. Pliny, The Elder (77AD) Natural History, Libri XIV Loeb Classical Library, London, England, 1945.
5. Licul R., Premužić D. (1979) Praktično vinogradarstvo i podrumarstvo. Nakladni Zavod Znanje, Zagreb, Croatia.
6. Fazinić N. (1981) Vineyards and wines of Croatia, Association for Development of Winegrowing, Zagreb, Croatia.
7. Tambača A. (1998) Vinogradarstvo i vinarstvo šibenskoga kraja kroz stoljeća. Matica Hrvatska Šibenik, Croatia.
8. Ravančić G. (2001) Život u krčmama srednjovjekovnog Dubrovnika. Hrvatski Institut za Povijest, Zagreb, Croatia.
9. Jelaska M. (1967) Ampelografija – Poljoprivredna enciklopedija. Jugoslavenski leksikografski zavod, Zagreb, Croatia.
10. Bulić S. (1949) Dalmatinska ampelografija. Poljoprivredni nakladni zavod, Zagreb, Croatia.
11. The official list of grape cultivars of Croatia, issued by the Ministry of Agriculture (N.N. 12/94 and N.N. 96/96)
12. Piljac J. (2002) Investigation of relatedness between Zinfandel and autochthonous Croatian grape varieties (*Vitis vinifera* L.) – Doctoral dissertation, Zagreb, Croatia.
13. Statistical information report for 2003. Central bureau of statistics, Republic of Croatia, Zagreb, Croatia.
14. Ričković M. (2002) Vinske pučke tradicije In: Milat V. (ed) Hrvatska vina i vinari. AGMAR, Zagreb, Croatia.
15. Žigić O. (1996) Vino i vinova loza u hrvatskim narodnim običajima i vjerovanjima, *Radovi hrvatskog društva folklorista*, p. 251-260.
16. Bašić Ž. (1999) Vina Dalmacije. Zadružni Savez Dalmacije "Zadrugar", Split, Croatia.
17. Mirošević N. (2002) Slijed sortimenta vinove loze u Hrvatskoj In: Milat V. (ed) Hrvatska vina i vinari. AGMAR, Zagreb, Croatia.
18. Sokolić I. (1997) Iskrice o vinu. Novi Vinodolski, Croatia.
19. Fazinić N., Fazinić M. (1996) Osvrt na vinogradarstvo otoka Korčule, *Godišnjak grada Korčule*, 1:217-224.

20. Bowers J. E., Boursiquot J. M., This P., Chu K., Johansson H., Meredith C. P. (1999) Historical genetics: The parentage of Chardonnay, Gamay, and other wine grapes of northeastern France, *Science*, 285:1562-1565.
21. Maleš P. (1998) Vino-prehrambeni proizvod primorskog krša. Vinoplod vinarija, Split, Croatia.
22. Mirošević N. (2002) Posuđeno iz povijesti za budućnost In: Milat V. (ed) Hrvatska vina i vinari, AGMAR, Zagreb, Croatia.
23. The Statute of the City of Dubrovnik, lib.2, c.18; lib.6, c.35 and 38.
24. Krivošić S. (1990) Stanovništvo Dubrovnika i demografske promjene u prošlosti, Zavod za povijesne znanosti HAZU, Dubrovnik, Croatia.
25. Kotruljević B. (1985) O trgovini i savršenom trgovcu (prijevod Ž. Muljačić) In: Radičević R. and Muljačić Ž. (eds) Djela znanosti Hrvatske, JAZU, Zagreb, Croatia, p. 117-240.
26. Nicolo di Gozze (1589) Governo della Famiglia, Appresso Francesco Ziletti, Venezia, Italy.
27. Nicolo di Gozze (1584) Discorsi sopra la Metheore d'Aristotele, Appresso Francesco Ziletti, Venezia, Italy.
28. Lučić J. (1991) Dubrovačke teme, *Matica hrvatska* no.154, Zagreb, Croatia.
29. The Statute of the City of Dubrovnik, lib.6, c.35.
30. Čremošnik (1933) Vinogradarstvo i vino u Dalmaciji srednjeg vijeka, *Glasnik Zemaljskog muzeja Bosne i Hercegovine*, 35:28.
31. Trummer X. F. (1841) Sistematische Classification und Beschreibung der im Herzogthume Steiermark vorkommenden Reben-Sorten, Graz, Austria.
32. Kraljević R. (1994) Vinogradarski slom i demografski rasap južne Hrvatske u osvit 20. stoljeća (Vinogradarstvo Dalmacije 1850.-1904.: uspon, procvat, klonuće). Književni Krug, Split, Croatia.
33. Kraljević R. (1982) Vinogradarska privreda Dalmacije (1850-1904) – Master's thesis, Mostar, Bosnia and Herzegovina.
34. Kraljević R. (1993) Uspon i pad vinogradarstva Dalmacije u drugoj polovici devetnaestog stoljeća, *Zadarska smotra*, 4-5:57-63.
35. *Narodni list* (1908) June 20, 50:1. Biankini's speech at the Emperor's palace in Vienna.
36. *Poljodjelski viestnik* (1893) August 16, 5:2.
37. *Narodni list* (1908) May 29, 43:544. Biankini's speach in Vienna: Brzopisna izvješća Dalmatinskog sabora (VI sjednica), Zadar.
38. Report from the Dalmatian parliament– Zadar, XXXVII, 1902. (VI. Meeting, July 8., 1902), p. 543 and 544.
39. Report from the Dalmatian parliament– Zadar, XXXVII, 1903. (I. Meeting, October 28., 1903), p. 441.
40. Mirošević, N. (1995) Vinogradarstvo Blata u rasponu vremena, *Zbornik-Duhovni i svjetovni obzori Blata na Korčuli*.
41. Milat V. (1995) Vino-život i put Blaćana, *Zbornik-Duhovni i svjetovni obzori Blata na Korčuli*.
42. Fazinić N., Fazinić M. (1983) Klimatske zone vinove loze u SR Hrvatskoj, *Jugoslavensko vinogradarstvo i vinarstvo*, 10-12:19-22.

43. Maleš P. (1997) Razvoj vinogradarstva i vinarstva na otocima, *Agronomski glasnik*, 2: 123-127.
44. Bowers J. E., Meredith C. P. (1997) The parentage of a classic wine grape, Cabernet Sauvignon, *Nature genetics*, 16:84-87.
45. Meredith C. P., Bowers J. E., Riaz S., Handley V., Bandman E. B., Dangl G. S. (1999) The identity and parentage of the variety known in California as Petite Sirah, *Am. J. Enol. Vitic.*, 50 (3):236-242.

Chapter 2: The story of Zinfandel

1. Kerridge G., Antcliff A. (1999) Wine grape varieties. CSIRO Publishing, Collingwood, Australia.
2. Verdegaal P.S., Rous C. (1995) Evaluation of Five Zinfandel Clones and one Primitivo Clone for Red Wine in the Lodi Appellation of California, Proceedings of the International Symposium on Clonal Selection, p.153.
3. Sall M.A, Teviotdale B.L., Savage S.D. (1981). Bunch rots, Grape Pest Management, Univ. of Calif. Div. Agric. Sci. Leaf., p. 51-56.
4. Winkler A.J., Cook, J.A., Kliewer W.M., Lider L.A. (1974) General Viticulture. University of California Press, Berkeley and Los Angeles.
5. Resource Guide to Zinfandel (1999) Zinfandel Advocates and Producers, Rough&Ready, California.
6. California Grape Acreage Report (2000) and California Grape Crush Report (2001). California Agricultural Statistics Service, Sacramento, California.
7. Tepsic A. (2002) TED Case Studies: Zinfandel, no. 662, January, 2002.
8. Voice of America interview / Jagoda Bush talked to Prof. Sullivan in December, 2002.
9. Senate resolution 132, 106th Congress, 1st session, July 1, 1999.
10. McGinty B. (1998) Strong Wine: The Life and Legend of Agoston Haraszthy. Stanford University Press, Stanford.
11. Schoenman T., Haraszthy A. (1979) Father of California Wine, Agoston Haraszthy: Including Grape Culture, Wines & Wine-Making. Capra Press, Santa Barbara, California.
12. Sullivan C.L. (2003) Zinfandel. A history of a grape and its wine. University of California Press, Berkeley and Los Angeles.
13. Sullivan, C.L. Zinfandel: A true vinifera. Vinif. Wine Growers J. 9:71-86 (1982).
14. Meredith C.P. (1996) Plavac mali: an academic view, *Wine Enthusiast*, October (1996), p.28.
15. *Wines and Vines*, April 1979, p.18.
16. Winkler A. J., Cook A. J., Kliewer W. M., Lider L. A. (1974) General Viticulture, University of California Press, Berkeley and Los Angeles.
17. Adams L.D. (1985) The Wines of America. McGraw-Hill, New York.
18. Mirošević N. and Meredith C.P. (2000) A Review of Research and Literature Related to the Origin and Identity of the Cultivars Plavac mali, Zinfandel and Primitivo, *Agriculturae Conspectus Scientificus*, 65(1):47-48.

19. Maleš P. (1993) Populacija Plavac (Plavac mali – Zinfandel – Primitivo), Vitagraf Rijeka, Rijeka, Croatia.
20. Trummer X. F. (1841) Sistematische Classification und Beschreibung der im Herzogthume Steiermark vorkommenden Reben-Sorten, Graz, Austria.
21. Bulić S. (1949) Dalmatinska ampelografija, Poljoprivredni nakladni zavod, Zagreb, Croatia.
22. Maleš P. (1998) Vino-prehrambeni proizvod primorskog krša. Vinoplod vinarija, Split, Croatia.
23. Bowers J. E., Meredith C. P. (1997) The parentage of a classic wine grape, Cabernet Sauvignon, *Nature genetics*, 16:84-87.
24. Meredith C. P., Bowers J. E., Riaz S., Handley V., Bandman E. B., Dangl G. S. (1999) The identity and parentage of the variety known in California as Petite Sirah, *Am. J. Enol. Vitic.*, 50 (3):236-242.
25. Bowers J. E., Boursiquot J. M., This P., Chu K., Johansson H., Meredith C. P. (1999) Historical genetics: The parentage of Chardonnay, Gamay, and other wine grapes of northeastern France, *Science*, 285:1562-1565.
26. Mirošević N., Pejić I., Maletić E., Piljac J., Meredith C.P. (2000), Relatedness of cultivars Plavac mali, Zinfandel and Primitivo (*Vitis vinifera* L.), *Agriculturae Conspectus Scientificus*, 65(1):21-25.
27. Fazinić N., Fazinić M. (1996) Osvrt na vinogradarstvo otoka Korčule, *Godišnjak grada Korčule*, 1:217-224.
28. Bašić Ž. (2001) Dalmatinska vina kroz stoljeća. Tiskara Kačić, Šibenik, Croatia.
29. Piljac J. (2002) Investigation of relatedness between Zinfandel and autochthonous Croatian grape varieties (*Vitis vinifera* L.) – Doctoral dissertation, Zagreb, Croatia.
30. Maletić E., Sefc K. M., Steinkellner H., Kontić J. K., Pejić I. (1999) Microsatellite variability in grapevine cultivars from different European regions and evaluation of assignment testing to assess the geographic origin of cultivars, *Theor. Appl. Genet.*, 100:498-505.
31. Sefc M. K., Lopes M.S., Lefort F., Botta R., Roubelakis-Angelakis K.A., Ibáñez J., Pejić I. Wagner H. W., Glössl J., Steinkellner H. (2000) Microsatellite variability in grapevine cultivars from different European regions and evaluation of assignment testing to assess the geographic origin of cultivars, *Theor. Appl. Genet.*, 100:498-505.
32. Jelaska M. (1967) Ampelografija – Poljoprivredna enciklopedija, Jugoslavenski leksikografski zavod, Zagreb, Croatia.
33. Wilfred W. (2002) Zinposium 2002: a major event, *Vineyard & Winery Management Magazine*, September/October '02.
34. Meredith C.P. (2003) Science as a Window into Wine History, American Academy of Arts and Sciences 1863rd Stated Meeting Report, November 2nd, 2002, Napa, California.
35. Voice of America interview / Jagoda Bush talked to Mary Lou Holt in December, 2002.

Chapter 3: Viticultural tradition of the Trogir-Kaštela surroundings

1. Babić I. (1978) Trogir, Nolit Beograd, Beograd, Yugoslavia, p. 6.
2. Stipčević A. (1991) Iliri, Školska knjiga, Zagreb, Croatia, p. 108.
3. Benić G. (2003) U Resniku se živjelo i prije 9000 godina, *Slobodna Dalmacija - Forum*, July 2003.
4. Petrić M. (1989) Amfore Jadrana. Logos, Split, Croatia, p. 14.
5. Kirigin B. (1996) Issa - grčki grad na Jadranu. Matica Hrvatske, Zagreb, Croatia, p. 135.
6. Bašić Ž. (2001) Dalmatinska vina kroz stoljeća. Tiskara Kačić, Šibenik, Croatia, p. 29.
7. Bašić Ž. (2001) Dalmatinska vina kroz stoljeća. Tiskara Kačić, Šibenik, Croatia, p. 30.
8. Ožanić S. (1955) Povijest dalmatinske poljoprivrede. Društvo agronoma NRH-podružnica Split, Split, Croatia, p. 149.
9. Ljubljanović S., Sokolić I. (1981) Iz prošlosti hrvatskog vinogradarstva, *Vjesnik*, August 27th and September 10th issues. In its original, this Latin proverb states: "Ubicumque Romanus vicit Romanus habitat."
10. Ožanić S. (1955) Povijest dalmatinske poljoprivrede. Društvo agronoma NRH-podružnica Split, Split, Croatia, p. 150.
11. Klaić V. (1985) Povijest Hrvata. Nakladni zavod Matice Hrvatske, Zagreb, Croatia, p. 46.
12. Klaić V. (1985) Povijest Hrvata. Nakladni zavod Matice Hrvatske, Zagreb, Croatia, p. 43.
13. Klaić V. (1985) Povijest Hrvata. Nakladni zavod Matice Hrvatske, Zagreb, Croatia, p. 86. Emperor Konstantin Portirogenet deals with the same topic in his publication "On ruling the empire" (De administrando imperio).
14. Ljubljanović S., Sokolić I. (1981) Iz prošlosti hrvatskog vinogradarstva, *Vjesnik*, August 27th and September 10th; issue no. 4.
15. Ožanić S. (1955) Povijest dalmatinske poljoprivrede. Društvo agronoma NRH-podružnica Split, Split, Croatia, p. 151.
16. Kečkemet D. (1978) Kaštel Sućurac. Mjesna Zajednica K. Sućurac, Split, Croatia, p. 48. Croatian rulers, just like most of the other rulers of their time, did not have a single residence; rather, they moved from court to court. Even ''the court' should not be considered equivalent to the palace; rather, it corresponded to a fortified estate. One such estate was established in Bijaći where the rulers/dukes issued official gift certificates.
17. Kečkemet D. (1978) Kaštel Sućurac. Mjesna Zajednica K. Sućurac, Split, Croatia, p. 50.
18. Klaić V. (1985) Povijest Hrvata. Nakladni zavod Matice Hrvaske, Zagreb, Croatia, p. 85.
19. Ljubljanović S., Sokolić I. (1981) Iz prošlosti hrvatskog vinogradarstva, *Vjesnik*, August 27th –September 10th, 1981, issue no. 5.
20. Andreis P. (1978) Povijest grada Trogira. Književni krug, Split, Croatia, p. 209.

21. Bašić Ž. (2001) Dalmatinska vina kroz stoljeća. Tiskara Kačić, Šibenik, Croatia, p. 56.
22. Klaić N. (1985) Povijest grada Trogira - Javni život grada i njegovi ljudi. Muzej grada Trogira V. 5, Trogir, Croatia, p. 276.
23. Bašić Ž. (2001) Dalmatinska vina kroz stoljeća. Tiskara Kačić, Šibenik, Croatia, p. 52.
24. The Statute of the city of Split - the medieval rights of Split (1985), Književni krug, Split, Croatia.
25. Novak G. (1978) Povijest Splita. Književni krug, Split, Croatia, p. 1362.
26. Ožanić S. (1955) Povijest dalmatinske poljoprivrede. Društvo agronoma NRH-podružnica Split, Split, Croatia, p. 23.
27. Ožanić S. (1955) Povijest dalmatinske poljoprivrede. Društvo agronoma NRH-podružnica Split, Split, Croatia, p. 26.
28. Čelebi E. (1967) Putopis: Odlomci o jugoslavenskim zemljama, Svijetlost, Sarajevo Bosnia-Herzegovina, p. 184.
29. Čelebi E. (1967) Putopis: Odlomci o jugoslavenskim zemljama, Svijetlost, Sarajevo, Bosnia-Herzegovina, p. 188.
30. Grgić I. (1962) Prva agrarna operacija na mletačkoj Novoj stečevini u Dalmaciji. (Naseljavanje novog stanovništva i razdioba zemlje na području Splita i Klisa 1672-73) Publication of the Split city museum, Split, Croatia, p. 6.
31. Ožanić S. (1955) Povijest dalmatinske poljoprivrede. Društvo agronoma NRH-podružnica Split, Split, Croatia, p. 164.
32. Ožanić S. (1955) Povijest dalmatinske poljoprivrede. Društvo agronoma NRH-podružnica Split, Split, Croatia, p. 154.
33. Bašić Ž. (2001) Dalmatinska vina kroz stoljeća. Tiskara Kačić, Šibenik, Croatia p. 85.
34. Kraljević R. (1994) Vinogradarski slom i demografski rasap južne Hrvatske u osvit 20. stoljeća (Vinogradarstvo Dalmacije 1850.-1904.: uspon, procvat, klonuće). Književni Krug, Split, Croatia, p. 30.
35. Kraljević R. (1994) Vinogradarski slom i demografski rasap južne Hrvatske u osvit 20. stoljeća (Vinogradarstvo Dalmacije 1850.-1904.: uspon, procvat, klonuće). Književni Krug, Split, Croatia, p. 135 (note 252).
36. Tartaglia M. (1899) Vinski vodič za Dalmaciju, Vinarska udružba za Dalmaciju, Split, Croatia, p. 45. There exist other sources about the harvest during this period, which cite different quantities; however the difference is insignificant. According to Tartaglia, 1888 marked the record harvest, and A. Agrario even mentions more than 2.000.000 hl of wine produced in that year.
37. Kraljević R. (1994) Vinogradarski slom i demografski rasap južne Hrvatske u osvit 20. stoljeća (Vinogradarstvo Dalmacije 1850.-1904.: uspon, procvat, klonuće). Književni Krug, Split, Croatia, p. 96.
38. Bašić Ž. (2001) Dalmatinska vina kroz stoljeća. Tiskara Kačić, Šibenik, Croatia, p.91.
39. Kraljević R. (1994) Vinogradarski slom i demografski rasap južne Hrvatske u osvit 20. stoljeća (Vinogradarstvo Dalmacije 1850.-1904.: uspon, procvat, klonuće). Književni Krug, Split, Croatia, p.80 and 156.

40. Kraljević R. (1994) Vinogradarski slom i demografski rasap južne Hrvatske u osvit 20. stoljeća (Vinogradarstvo Dalmacije 1850.-1904.: uspon, procvat, klonuće). Književni Krug, Split, Croatia, p. 180.
41. Kraljević R. (1994) Vinogradarski slom i demografski rasap južne Hrvatske u osvit 20. stoljeća (Vinogradarstvo Dalmacije 1850.-1904.: uspon, procvat, klonuće). Književni Krug, Split, Croatia, p. 199.
42. Tambača A. (1998) Vinogradarstvo i vinarstvo šibenskoga kraja kroz stoljeća. Matica Hrvatska Šibenik, Croatia, p. 198.
43. Kraljević R. (1994) Vinogradarski slom i demografski rasap južne Hrvatske u osvit 20. stoljeća (Vinogradarstvo Dalmacije 1850.-1904.: uspon, procvat, klonuće). Književni Krug, Split, Croatia, p. 175.
44. Kraljević R. (1994) Vinogradarski slom i demografski rasap južne Hrvatske u osvit 20. stoljeća (Vinogradarstvo Dalmacije 1850.-1904.: uspon, procvat, klonuće). Književni Krug, Split, Croatia, p. 152.
45. Bašić Ž. (2001) Dalmatinska vina kroz stoljeća. Tiskara Kačić, Šibenik, Croatia, p.110.
46. Bašić Ž. (2001) Dalmatinska vina kroz stoljeća. Tiskara Kačić, Šibenik, Croatia, p.117
47. Bašić Ž. (2001) Dalmatinska vina kroz stoljeća. Tiskara Kačić, Šibenik, Croatia, p.118.
48. Ožanić S. (1955) Povijest dalmatinske poljoprivrede. Split, Croatia, p. 402.
49. Bašić Ž. (2001) Dalmatinska vina kroz stoljeća. Tiskara Kačić, Šibenik, Croatia, p.151.
50. Ožanić S. (1955) Povijest dalmatinske poljoprivrede. Društvo agronoma NRH-podružnica Split, Split, Croatia, p. 165.
51. Bašić Ž. (2001) Dalmatinska vina kroz stoljeća. Tiskara Kačić, Šibenik, Croatia, p. 153.
52. Bašić Ž. (2001) Dalmatinska vina kroz stoljeća. Tiskara Kačić, Šibenik, Croatia, p. 156.
53. Several authors (Maleš, Tomić, Pezo, Rumora and Gazari) (1984) Zaštita geografskog porijekla kvalitetnog vina Kaštelet - Institute for Adriatic Crops, Split, cited in: Bašić Ž. Vina Dalmacije (1999), Zadružni Savez Dalmacije "Zadrugar"', Split, Croatia, p. 42.

Chapter 4: A special kind of a wine story

1. Pejić I., Maletić E., Kontić J.K, Kozina B., Mirošević N. (2000) Diversity of autochthonous grapevine genotypes in Croatia, VIIth International symposium on grapevine genetics and breeding, Proceedings of the Seventh International Symposium on Grapevine Genetics and Breeding, Montpellier, July 1998. *Acta Horticulturae*, 528:69-73.
2. Tambača A. (1998) Vinogradarstvo i vinarstvo šibenskoga kraja kroz stoljeća. Matica Hrvatska Šibenik, Croatia.

3. Bašić Ž. (2001) Dalmatinska vina kroz stoljeća, Tiskara Kačić, Šibenik, Croatia.
4. Maleš P. (1998) Vino-prehrambeni proizvod primorskog krša. Vinoplod vinarija, Split, Croatia.
5. Kiš D. (2002) Primoštenski vinogradi, *Croatia - a traveler's magazine*, Autumn 2002.

SELECT BIBLIOGRAPHY

AMERINE A., MAYNARD and SINGLETON L., VERNON – Wine: An introduction. University of California Press: Berkeley - Los Angeles - London 1965.
ANDREIS, PAVAO – Povijest grada Trogira. Književni Krug, Split 1978.
BAŠIĆ, ŽELIMIR – Dalmatinska vina kroz stoljeća. Tiskara Malenica, Šibenik 2002.
BAŠIĆ, ŽELIMIR – Vina Dalmacije. Zadružni savez Dalmacije – Mediteranska poljoprivredna biblioteka, Split 1999.
BABIĆ, IVAN - Trogir. Nolit, Beograd 1978.
BENIĆ, GORDANA – U Resniku se živjelo i prije 9000 godina. Slobodna Dalmacija-Forum, Split, July 2003.
BULIĆ, STJEPAN – Dalmatinska ampelografija. Poljoprivredni nakladni zavod, Zagreb 1949.
ČELEBI, EVLIJA – Putopis (Seyahatname): Odlomci o jugoslavenskim zemljama. Translation by Hazim Šabanović, Svjetlost, Sarajevo 1967.
DELALLE, IVAN – Trogir-vodič. Općinski turistički savez, Trogir 1966.
JELASKA, MARCEL – Osnovni problemi vinogradarske proizvodnje. Zbornik društva inženjera i tehničara, Split 1958.
KEČKEMET, DUŠKO – Kaštel Sućurac, Mjesna zajednica K. Sućurac, Split 1978.
KIRIGIN, BRANKO – Issa – grčki grad na Jadranu, Matica Hrvatska, Zagreb 1996.
KLAIĆ, NADA – Povijest grada Trogira - Javni život grada i njegovi ljudi. Museum of the city of Trogir, volume 5, Trogir 1985.
KLAIĆ VJEKOSLAV – Povijest Hrvata. Nakladni zavod Matice Hrvatske, Zagreb 1985.
KRALJEVIĆ, RUDOLF – Vinogradarski slom i demografski raspad južne Hrvatske u osvit 20 stoljeća, Split 1994.
LICUL, RANKO and PREMUŽIĆ, DUBRAVKA – Praktično vinogradarstvo i podrumarstvo, Nakladni zavod Znanje, Zagreb 1982.
MALEŠ, PETAR – Plavac. Institut za jadranske kulture i melioraciju krša, Split 1981.
MALEŠ, PETAR – Vinogradarstvo i vina Dalmacije. Institut za jadranske kulture i melioraciju krša, Split 1985.
MALEŠ, PETAR – Vino - prehrambeni proizvod primorskog krša. Vinoplod, Split 1998.
MIROŠEVIĆ, NIKOLA and TURKOVIĆ ZDENKO - Ampelografski atlas. Golden marketing-Tehnička knjiga, Zagreb 2003.
NOVAK, GRGA – Povijest Splita. Književni Krug, Split 1978.
OMAŠIĆ, VJEKO Sedam Kaštela – Iz kulturnog nasljeđa Kaštela (catalogue). Zavod za zaštitu spomenika kulture, Split 1988.

OŽANIĆ, STANKO – Povijest dalmatinske poljoprivrede. (Prepared and published by A. Tambača and D. Morović according to the original "Prilozi za povijest poljoprivrede Dalmacije") Društvo agronoma NRH-područnica Split, Split 1955.

PERIČIĆ ŠIME – Gospodarske prilike u Dalmaciji od 1797. do 1848. Književni Krug, Split 1993.

SOKOLIĆ, IVAN – Iskrice o vinu. Personal publishing, Novi Vinodolski 1997.

SOKOLIĆ, IVAN - Prvi hrvatski vinogradarsko – vinarski leksikon. Personal publishing, Novi Vinodolski 1998.

STIPČEVIĆ, ALEKSANDAR – Iliri. Školska knjiga, Zagreb 1991.

SULLIVAN L., CHARLES – Zinfandel: A history of a grape and its wine. University of California Press: Berkeley - Los Angeles - London 2003.

TAMBAČA, ANDRIJA – Vinogradarstvo i vinarstvo šibenskog kraja kroz stoljeća. Matica Hrvatska, Šibenik 1998.

TARTAGLIA, MIHO – Vinski vodič za Dalmaciju. Vinarska udružba za Dalmaciju, Brzotisak "Narodni tisak", Split 1899.

ZRNČIĆ, MIROSLAV – Podrumarstvo. Globus, Zagreb 1993.

SEVERAL AUTHORS – (Winkler et al) General viticulture. University of California Press: Berkeley & Los Angeles 1974.

SEVERAL AUTHORS – Hrvatska vina i vinari. AGMAR, Zagreb 2002.

SEVERAL AUTHORS – Vina Dalmacije. Poslovno udruženje dalmatinskih vinara, Split 1964.

SEVERAL AUTHORS – Enciklopedija Leksikografskog zavoda. Leksikografski zavod, Zagreb 1966.

SEVERAL AUTHORS – (Sokolić et al) Zlatna knjiga o vinu. Otokar Keršovani, Rijeka 1976.

SEVERAL AUTHORS – (Kirigin et al) Otok Vis u helenističko doba (catalogue). Arheološki muzej, Split 1983.

FACTS ABOUT CROATIA AND WINE

Area: 56,538 square kilometers.

Inhabitants: 4,800,000.

Nationality: 78% Croatians.

Religion: 76.5% Roman-catholic.

Capital: Zagreb, 1,200,000 residents.

Larger cities and ports: Osijek, Split, Rijeka, Dubrovnik.

Length of the coastal belt: 5,790 km, 1,778 km of which is mainland coast and 4,012 km island coast.

Number of islands: 1185, 66 of which are inhabited.

Larger islands: Krk, Cres, Brač, Hvar, Pag, Korčula, Dugi otok, Mljet, Vis, Rab, Lošinj.

Number of national parks: eight, four of which are in the mountains – Risnjak, Sjeverni Velebit, Paklenica and Plitvička jezera.

Nature reserves, parks and monuments: 1,111,950 acres with 44 protected rare plant species and 381 animal species.

Area under vineyards: according to official statistics 143,318 acres, 32,864 acres of which have been registered at the National institute for viticulture and enology.

Area under vineyards according to viticultural regions: 35% northwestern Croatia, 35% Dalmatia, 15% Slavonija and Podunavlje, 15% Istria and northern coastal belt.

Number of cultivars: total number aproximated at 230, 143 included in the official Croatian cultivar list - more than 80 are considered to be native.

Most important autochthonous varieties: Graševina (25.5%), Moslavac, Malvazija (11.7%), Pošip, Rukatac, Bogdanuša, Grk, Vugava, Plavac mali (10.6%), Babić, Teran, Refošk.

Wine production: according to official statistics slightly over 2,000,000 hl, 700,000 hl of which is wine with controlled appelation label. White wine accounts for about 2/3 of total wines produced, 1/3 are reds, and rosés are negligible.

Number of labels: more than 1000, 900 of which carry the appelation label.

Percentages of wine types: 5% top quality wines with controlled appelation, 55% quality wines with controlled appelation.

Wine producers: 350 produce wine with controlled appelation; there are about 30 large or intermediate-size wine cellars and about 30 cooperatives, the rest are individual producers.

The largest wine producers and cellars: Badel 1862 – Zagreb, Dalmacijavino – Split, Kutjevački podrum – Kutjevo, Đakovačka vina – Đakovo, Agrolaguna – Poreč, Ivex – Rijeka, Vinoplod – Šibenik, Iločki podrum – Ilok.

The largest individual producers: Ivan Enjingi – Slavonia (111 acres), Ravalico family – Istrian peninsula (64 acres).

The most famous label: Dingač (made from Plavac mali grapes) – also, the first wine with protected geographical origin; 47 ha of Dingač accounts for about 1/10 of the total areas under Plavac mali variety on the Pelješac peninsula.

Annual wine consumption: according to statistics about 50 L per person, same indices point to a lower number of about 25 L per person.

BIOGRAPHY

Born in 1976, in Zagreb, to a couple of well-known scientists, who dedicated years of research to biomedicine, Jasenka Piljac experienced the beauty and hardships of two worlds, Croatia and California. During her seven-year-long biochemistry and molecular biology studies in California, the author found that the one thing common to scientists in both worlds is the eagerness for scientific truth.

In the laboratory of Professor Carole Meredith at UC Davis, where prestigious grape genetics studies were conducted, Jasenka Piljac learned the steps involved in practical, real science. Partly due to patriotic reasons and partly due to scientific curiosity, her interests in grape genetics were further deepened during the search for Croatian origins of the famous Californian Zinfandel grape. Along with a Ph.D. in molecular biology, and numerous scientific and professional publications, the result of the author's infatuation with Dalmatia, tradition and grapes is this comprehensive publication – the truth about Zinfandel.

The author's wish to eventually return to Croatia and continue with grape and wine research was granted in 1998. She is currently working on several grape-related projects at the multidisciplinary research institute "Ruđer Bošković" in Zagreb.